DEAD HEAT

DEAD HEAT

The Race Against the Greenhouse Effect

Michael Oppenheimer

Robert H. Boyle

I.B.TAURIS & Co Ltd

Publishers

London · New York

Published in 1990 by
I.B. Tauris & Co Ltd
110 Gloucester Avenue
London NW1 8JA

British Library Cataloguing in Publication Data

Oppenheimer, Michael
  Dead heat: the race against the greenhouse effect.
  1. Climate.   Effects of carbon dioxide
  I. Title   II. Boyle, Robert H.
  551.6

  ISBN 1–85043–241–4

Printed in Great Britain by
Redwood Press Limited, Melksham, Wiltshire

To Leonie,
for absolutely everything,
and to Katya,
Я Тебя ЛЮбЛЮ

Human history becomes more and more a race between education and catastrophe.
— H. G. Wells

CONTENTS

Contents

ACKNOWLEDGMENTS

MANY PEOPLE HAVE CONTRIBUTED FREELY OF THEIR TIME IN helping us assemble this book. We would like to thank in particular Lauri Adams, Aramando Arredondo, James Atz, Richard Ayres, Terry Backer, Bob Brown, Debbie Cohen, John Cronin, Jane Cushman, Sally Dorst, Daniel J. Dudek, Harvey Flad, John Fry, Gordon Goodman, Joshua Haimson, Ria and Charles E. Harris, Jr., Jill Jäger, Charles Komanoff, Peter Lehman, Allan Margolin, Peter Miller, Sidney Morgenbesser, Richard Nelson, Joel O'Connor, Joan Ogden, Shirley Oppenheimer, Rafe Pomerance, Bruce Rich, James P. Rod, Robert Stavins, William R. Schell, Stephen Schwartzman, Robert Sullivan, James T. B. Tripp, Joseph Vecchione, Francis J. Voyticky, Thomas Whyatt, Robert Williams, Steven Wofsy, George Woodwell, and Alexander Zagoreos. RHB would also like to thank the staffs of the Butterfield Memorial Library, Cold Spring, N.Y., and the Alice and Hamilton Fish Library, Garrison, N.Y., and most of all, Ophelia Layug, Stephanie Boyle Mays, and Kathryn Belous-Boyle.

Leonie Haimson not only provided MO with unlimited amounts of love, encouragement, and insight, without which he could not have proceeded with this book, but also edited the manuscript thoroughly at two different stages, bringing to it a sorely needed coherence. MO might not have begun the project at all were it not for Peter Kaplan's persistence in pressing the idea upon him, year in and year out, and then helping him through the thicket of the publishing world. Arthur Wang also provided invaluable guidance at an early stage. Leslie Daniels and Verne Dawson supplied us with their unmatched research skills. Professors M. Hoque and R. Khandoker were extremely generous in guiding MO about Bangladesh. Bill Newlin provided a keen edit of the manuscript. We thank the Environmental Defense Fund and its executive director Frederic D. Krupp for their extreme generosity in allowing MO the time to complete the book. Finally, Georgia Pease suffered two overbearing egos patiently while typing the manuscript time after time and keeping MO's professional existence afloat simultaneously. To her we are most indebted.

DEAD HEAT

PROLOGUE

HUMANITY IS HURTLING TOWARD A PRECIPICE. LEFT UNCHECKED, the emissions of various gases, particularly carbon dioxide from fossil-fuel combustion and deforestation, are likely to alter the Earth's climate so rapidly and so thoroughly as to destroy much of the natural world and turn the world that we call civilization upside down. But such an outcome is not inevitable. The shape of the future remains ours to choose.

For the past 200 years, the atmospheric level of carbon dioxide and other greenhouse gases has been climbing as industrialization has moved apace. A 100-year-old scientific theory, backed up by modern computer analysis, indicates that this growth in gases is trapping heat and warming the Earth and, indeed, the planet is now about one degree Fahrenheit warmer than it was in the late nineteenth century. One symptom of the trend is that, as of 1989, five of the past nine years have been the hottest in the 130-year record of global temperature measurement.

Because climate also varies in tune with natural factors, the Earth's temperature does not move uniformly but changes in fits and starts. For example, an increase in emissions of volcanic dust (which reflects sunlight), a tiny fading of the sun's intensity, or perhaps a random flicker in the climate halted the warming between 1940 and 1970

1

and actually caused the Northern Hemisphere to cool slightly. Natural fluctuations can cause modest warming, too, and as a consequence, there remains uncertainty over how much of the one-degree trend is attributable to the greenhouse effect.

Even so, the continuing growth of these gases in the atmosphere is threatening to overwhelm all such natural variations. During the next 75 years, enough gas could accumulate to heat the Earth by more than eight degrees, nearly as much as it warmed during the last world-shaping climate change as the Pleistocene glaciers, which had been two miles thick at their maximum, began their retreat from North America and Eurasia when the Ice Age drew to a close 15,000 years ago. That warming stretched over 5,000 years; we now face equivalent change, compressed into a mere century.

The full impact of greenhouse gases will not be felt for decades after their emission, and already, it may be too late to avoid an additional two-degree warming. By 2025, the Earth could be three degrees warmer than today, a point which marks the boundary of a climatic no man's land. The planet has not been hotter for more than 200,000 years, long before modern humans evolved. Scientists have only the dimmest notion of how climate behaved and how nature responded then. A five-degree jump could occur by 2050, within the lifetime of today's children, making the Earth hotter than it has been for more than 4 million years, which is before the time of the earliest known biped, *Australopithecus afarensis*, Johanson's Lucy. By the end of the twenty-first century, the world could be warmer than at any time since the dinosaurs expired 65 million years ago, when Antarctica was ice-free and alligators roamed far into the latitudes of what is now Canada.

Climate changes of this order cannot help but remake the face of the earth. As glaciers melt and the seas heat and expand, the level of the ocean will rise, reconfiguring coastlines worldwide. Plants and animals are adapted to particular climate zones, and as the Earth warms, they will attempt to shift northward. The survival of forests will depend on seeds reaching appropriate regions where new forests can take hold before zones shift again. But climate zones may move north 400 miles in the next 100 years, much faster than seeds were able to migrate when the Pleistocene glaciers retreated. During the recovery from the last ice age, at least thirty-two genera of mammals became extinct, and the potential loss of plant and animal species from the impending global warming appears overwhelming. Today, cities, towns, factories, rail lines, and other "infrastructure" will impede or block animals migrating north as well as seeds carried by birds and the wind. With their progeny outraced by the fast-moving climate, the old forests will succumb to heat and drought and eventually burn to the ground. Many species will simply disappear, forever.

In the face of shifting conditions, humans may also be racing to maintain food supplies and surely will be struggling to protect buildings, beaches, land, and people from the rising sea. Some areas will be abandoned altogether, others preserved at great expense. The global economy will gradually feel the strain of the warming as farms, industries, and tens of millions of people are forced to migrate and then move again; as natural resources are stretched to the limit and finally are exhausted; as conflicts erupt in the struggle over what remains.

Averting this bleak scenario means slowing emissions of the greenhouse gases, and there is nothing to be gained by waiting. Once the unbearable consequences are upon us, more warming will inevitably result from the delayed action of gases already emitted. So the answer to the question of when to move against fossil fuels, such as coal and oil, is obvious: we should have already begun.

Fortunately, fossil fuels are neither the sine qua non for an advanced living standard nor the key to prosperity for the world's poor. Solar energy in its various forms can provide a technically adequate and environmentally superior alternative. Over the next few decades, its cost could also become competitive, gradually shrinking the demand for fossil fuels. Such transitions in energy sources have occurred regularly since the Industrial Revolution. A different source has been part and parcel of each of the four past industrial periods, and a transition to a fifth era is well under way. The challenge is to speed the transformation and channel the economy toward the quickest and most complete substitution for fossil fuels before the world is overwhelmed by climatic change.

A small group of economists has recently explored the relationship between technological change, such as the introduction of new energy sources, and economic development. Their work proves to be particularly useful for understanding the transition to the next industrial era, which they call the fifth wave, and we shall borrow their use of this term. The fifth wave will see the replacement of mass production and heavy industry by a computer-controlled, information-communications economy. Semiconductors are the stuff of which both computer chips and solar cells are made. The merging of these technologies with lightweight but superstrong materials will first make fossil-fuel use more efficient, and then replace it altogether. Solar cells aren't new, but their thirty-five-year existence is a blink of an eye in technological time; historically, it has taken twice as long for humans to develop their inventions fully and another sixty to eighty years to exhaust their potential.

There is both good news and bad news in this assessment. The good news is that replacing fossil fuels need not be viewed as an expensive and wrenching distortion of the normal rhythms of the global economy. Quite the opposite. To the extent that the shift is encouraged, movement is accelerated toward an era that promises economic growth based on a resource that will not soon be exhausted—the sun. The age of oil, in particular, will not last another century, and substitutes would have to be found even if the greenhouse effect were not a problem. In the long term, then, what's good for the environment is good for General Motors (or at least for Toyota).

The bad news is that, left to its own devices, the global economy would not make the transition nearly fast enough to avoid environmental catastrophe. And because the transformations needed to take full advantage of advances in energy and transportation are social as well as technical, the fossil-fuel economy can't be scrapped overnight. For example, motor vehicles consume one quarter of the energy used each year in the EEC countries and produce an equivalent percentage of carbon dioxide emissions. But cars and trucks cannot simply be abandoned, because housing, manufacturing, and employment—an entire way of life—depend on them. Information may be more valuable than hard goods in the future and so fiber optics may be more in demand than sixteen-wheelers, and home computers may be more significant than commuters, but implementing this change will entail restructuring society considerably. Cars could be powered by electricity, or hydrogen fuel made from

4

solar energy, but the switch from gasoline will become practical only with the advent of lightweight materials and shorter commuting distances. In other words, no solution can be dropped unconditionally into the current framework. The whole social-technical pattern must change together.

Furthermore, there are historical circumstances that could actually accelerate fossil-fuel use in the near term. The biggest challenge is posed by the continued economic and population expansion of such Third World countries as China and India. Even if the developed countries shift away from fossil fuels, Third World growth could continue to be based on coal and oil, in which case there would be little hope of restraining global warming short of a calamity. Limiting economic growth is not an option, for this would only succeed in trapping the Third World in a cycle of poverty that would eventually disrupt the international order just as surely as an environmental disaster. The resolution of this conflict is to be found in the free transfer of new, solar technologies to the less developed world, as well as the intensified encouragement of programs that would lead to the ultimate stabilization of their populations. If this leapfrog over heavy industry can be achieved, the Third World could actually benefit from the greenhouse quandary: with sun to spare, many could export stored solar energy the way some now export oil.

The only tool at hand for speeding a global transition to the fifth wave is government; and government, particularly in the United States, has fallen into a period of self doubt and financial rigor mortis that has paralyzed it even when confronted with problems of much lesser magnitude. Accordingly, there is good reason for skepticism. Yet there is also reason for optimism. The Cold War is ending and political leaders have lost their bearings. This uncertainty, along with the dawning appreciation for the fragility of the planet which followed the discovery of the ozone hole, has permitted the environmental movement to place its agenda at the top of the list. Already, in Europe the Greens have moved quickly into the void and promise to exert more and more influence in the European Parliament as 1992 approaches. As a result, the leaders of Great Britain, West Germany, France, and other nations are scrambling to develop environmental credentials.

An opportunity has emerged to shift from the military-industrial complex to a political and economic revitalization that is environmentally driven. A few European countries are already considering ways to cut their fossil-fuel dependence and simultaneously profit from the new competitive opportunities unearthed in the process. The United States has responded more slowly; as the financial commitment to solar energy in West Germany and Japan is growing, ours is declining. The environment could provide us with a rallying point, an organizing principle for making economic and political choices, much the way the Cold War did for the past forty years. But given our current course, it is not difficult to imagine a future where neither the technical nor political impetus behind the greenhouse solution will come from the United States, because we have invested little in it and as a result have nothing to sell.

If greenhouse gases keep growing, what will the planet be like sixty years from now, no further in the future than the Great Depression is in the past? The blistering summer of 1988 provides a useful analogue. Heat waves baked much of the country and drought cut grain production by more than a third, sent Yellowstone up in flames with smoke that generated its own lightning, and lowered the level of the Mississippi to the point where barges became mired in the mud bottom. The outlines of an answer also emerge from computer simulations of climate. The evolution of future temperatures described in this book (absent action to limit the gases) is based on one projection by the Goddard Institute for Space Studies, using a so-called "business-as-usual" scenario, which assumes the continuation of present trends in growth of greenhouse gas emissions. Unless otherwise specified, numbers refer to estimated increases above today's global mean temperature (in degrees Fahrenheit).*

A healthy respect for scientific uncertainties is in order. As a rule of thumb, the various projections of future climate are such that a particular global mean temperature might be 20 percent higher or 50 percent lower than the Goddard estimate for any future period, given the same emissions. Warming at the lower rate would present

---

*Global mean temperature is calculated by averaging measurements from thousands of monitoring stations worldwide.

fewer difficulties than warming according to the higher estimates, but even slow warming would move climate outside the realm of experience of *Homo sapiens sapiens*, the modern human species, during the next century. Uncertainties increase when smaller areas than the world as a whole (like the United States) are considered. Furthermore, phenomena known as feedbacks, which affect the response of climate to greenhouse gases, are not fully captured by any simulation. These could ameliorate the warming, or just as likely, make it considerably more severe than expected.

Bigger problems arise in projecting changes in storm patterns and precipitation, or runoff into rivers and reservoirs that is ultimately utilized by agriculture, industry, and as drinking water. Higher temperatures mean faster evaporation of ocean water, and more precipitation as a result. But moisture will evaporate faster from soils, too, and the balance between these two effects will vary from place to place. The consequences are difficult to calculate. There is no firm consensus, even for an area the size of the Great Plains breadbasket. For example, a model developed by scientists at the Geophysical Fluid Dynamics Laboratory in Princeton projects a drying of soils in the critical summer growing season throughout America. On the other hand, the Goddard computer simulation shows a drying of soils over large parts of the Plains, the Southeast, and the Midwest, but increased moisture along the northern tier. Most likely there will be a tendency for places that are already dry at certain times of year to become drier, and places that are already wet to become wetter; and conditions everywhere will change.

In the chapter that follows, we supplement this information with fossil records that indicate how fast plants and animals migrated when the last ice age ended, along with some educated guesses about human responses in the future, in order to develop a plausible scenario for the next sixty years. The composite picture itself is fictive, and no computer analysis could either prove or disprove it. But the individual events are distilled from scientific understanding, as described in the endnotes. Some may occur at a later date than we have placed them, some sooner. Eventually, all could occur somewhere, sometime, if we fail to act.

CHAPTER 1

The End

IMAGINE . . . THE YEAR IS 2050. GLOBAL WARMING HAS REMADE THE face of the Earth, but no continent has been ravaged more than North America. Even before the greenhouse age, its landmass experienced climate's greatest excesses. The Pacific Ocean on the west, the Atlantic Ocean on the east, the cold waters of the Arctic to the north, and the warm waters of the tropics to the south provided sharp temperature contrasts and ample moisture, while the Rocky Mountains and the Appalachians drove air currents upward and gravity pulled them down. As a result, hurricanes, blizzards, thunderstorms, bitter cold spells, tornadoes (almost unique to the continent), drenching rains, and blistering heat waves had always played out a life-and-death drama across the land. These extremes slowly and at first imperceptibly melded with the changing climate. The year 1988–1989 saw drought in the Midwest followed by erratic shifts in the jet stream that brought record low temperatures to Alaska and snow to Los Angeles. What were once anomalies have become a matter of course.

All debate about global warming ended in 1998 after a four-year drought desolated the heartlands of North America and Eurasia. In 1995, food riots in Kiev, Cherkassy, and Odessa sparked a new resurgence of Ukranian nationalism, prompting the neo-Stalinists, who had overthrown Mikhail Gorbachev, to start a brutal repression that made even the Chinese call for UN sanctions. In the plains states, from Iowa to eastern Colorado, south to Texas and north to South Dakota, the age of the family farm finally came to an end, and the sturdy freeholders, long seen as an anchor of U.S. democracy, dispersed. Some signed on with the agribusiness conglomerates that bought up land and lobbied Congress for pipelines to the Great Lakes before the water levels there fell, too, while others sought to start over in Alberta, Saskatchewan, or Manitoba. But most of them, joined in the next two decades by a swelling trek of tens of thousands of others from bankrupt farm and ranch communities, looked for jobs in the cities of the upper Midwest and Canada. Duluth now bulges with 1.3 million people; Edmonton, 6 million; Toronto, 11 million. Forty years have passed since farmers in John Deere hats gathered for morning coffee at the Rocket Inn in Roland, Iowa, to chew the fat about the weather, the cost of machinery, and how the Cubs were doing, and thirty years have gone by since the high school band last played in Valentine, Nebraska, before the Badgers' big game against Ainsworth. They were among the first communities to empty, the precursors of thousands of ghost towns that stipple the plains from Colorado to Indiana.

To many Americans and Canadians, the greenhouse signal literally became visible during the last two weeks of October of 1996, when winds that seemed to roar without respite gathered a "black blizzard" of prairie topsoil that darkened the skies of sixteen states and the Canadian Maritimes. The dust penetrated the lungs of cattle, stopped traffic on interstates, stripped paint from houses, and shut down computers. People put on goggles and covered their noses and mouths with wet handkerchiefs. They stapled plastic sheets over windows and doors but still the dust seeped through. Analysis disclosed that soil from Dalhart in the Texas Panhandle landed as far away as Halifax, Nova Scotia. In place of the soil, the winds left only the heavy sands that now bury parts of the western plains under drifting dunes.

As the sands advanced, the Platte River in Nebraska dried up, and the wetlands of the Cheyenne Bottoms in Kansas withered away completely. The several million waterfowl and wading birds—geese, ducks, swans, shorebirds, and cranes—that stopped there in spring on their way north from South and Central America were devastated. The first to expire were the shorebirds, gone by the turn of the century. They could not find food to add enough fat to fuel a flight of even 200 miles, let alone the 1,500 miles they needed to reach their Arctic nesting grounds.

Before the detection of the greenhouse signal, some agronomists optimistically contended that crops would thrive if planted further north. They had in mind the so-called carbon-3 plants—wheat, oats, barley, and sunflowers—that are more efficient photosynthesizers than corn, sorghum, and other carbon-4 plants. In theory, the C3 plants would benefit from the extra carbon dioxide in the air. But then, so did many weed species. Even in areas where weed control worked and soils proved suitable, some C3 plants utilized the carbon dioxide to add leaf growth rather than food and fiber, and populations of cutworms, sunflower moths, and a hitherto rare species of leafhoppers exploded. American grain reserves hit zero with the 1990s drought, but shortages were staved off by increased herbicide and pesticide use, and by purchases from Canada, which accelerated forest cutting to produce more crop land. Still, reserves stayed low, and when the Indian monsoon failed in 2005, no one could help avoid the famine. The improving yields in northern Russia had to be used to offset losses in the Ukraine instead.

Political action in Washington was stymied until the first decade of this century because California and some other populous states had been spared. Then the patterns began to change. For a time, the West Coast received as much winter precipitation as before, but it came as rain, not snow. In California, what little snow fell on the Sierra Nevada soon melted. Reservoirs lacked the capacity to store the sudden runoff that scoured the slopes.

The ensuing water shortage cut generation of hydroelectric power at the peak of summer demand for air conditioning and refrigeration, which heightened the long-standing battle over water allotments between urban residents and growers. Already upset at watching their beach houses at Malibu slide into the rising sea, sweltering Angelenos joined forces with other city dwellers and tried to strong-arm the legislature and Congress into barring irrigation of alfalfa, cotton, and rice, the most water-consumptive crops in the state. The effort failed temporarily, but a new crisis erupted a decade ago during a record four-year drought. Under political pressure from growers, managers released too much water during the first year of the drought, and subsequently little remained in reserve. Yields declined sharply, and the restrictions on irrigation were finally imposed. Back east, prices for vegetables skyrocketed, and colored soy substitutes from Canada replaced them. In Mexico, police began rounding up illegal American migrants working the fields around Monterrey.

With no spring runoff, and stressed by heat and smog, forests in the Sierra Nevada and the hills and mountains around Los Angeles began to die, and, as elsewhere in the country, the dead trees fueled catastrophic fires that were impossible to contain. For six months, no one could see across the Grand Canyon. The smoke and the heat and the smog drove people out of Phoenix and Tucson and back to the north—people whose families had originally moved there because of the erstwhile clean air.

In northern California, low water levels and high temperatures deoxygenated Tule Lake, inducing epizootics of botulism that eventually killed off the immense flocks of ducks and geese that had made Tule the greatest single gathering ground in the world for migratory waterfowl. As late as the fall of 1960, 10 million birds from Canada, Alaska, and Siberia had stopped by the lake. By 1990, the overall number was down to 4 million, largely as the result of the drainage of wetlands for agriculture, particularly in Canada; and, as farming shifted farther north in the prairie provinces, more wetlands were drained and waterfowl continued to decrease in numbers.

During the heat wave of October 2006, disaster struck at Tule. Birds died after eating dead invertebrates, mainly fairy shrimp and midge larvae, containing the fatal poison produced by the anaerobic bacterium *Clostridium botulinum*. Flies then deposited their eggs in the dead birds, and the developing maggots absorbed the poison. Other birds gorged themselves on the maggots in the dead birds, and the "maggot cycle" went wild. All told, the body count made by the U.S. Fish and Wildlife Service showed that of the 300,000 birds that stopped at Tule that fall, 270,000 perished. Last year, 2049, no waterfowl stopped at Tule. Once hailed in song as the Golden State, California has become the State of Death.

Meanwhile, forest death crept up the East Coast from Virginia to Maine. The last red spruce in the Green Mountains of Vermont died in 2010, and within twenty years vast stands of skeletonized trees—maples, ashes, pines, beeches, hickories, hemlocks, birches, oaks, sycamores—ran from New England south to Pennsylvania, sporadically erupting into fire and smoke that choked the Northeast. What trees were left had to struggle to survive. Neither the wind nor the diminishing number of birds and squirrels could disperse seeds fast enough to permit colonization to the north, and the federal government entered into a cooperative tree-bank planting program with Canada. The thin, impoverished, coarsely textured soils of the Canadian Shield in Ontario and Quebec prevented the establishment of many tree species, and invading pests, most prominently the gypsy moth, defoliated or otherwise damaged the new plantations. No matter: tree planting proved fruitless anyway, because the big canopy trees could not grow fast enough to catch up with the climate in which they were to mature. The temperate zone was pushing northward too quickly.

In 1993 and 1996, heat waves struck the Southeast, cutting corn and soybean production by 50 percent despite sufficient rainfall. Timber yields also declined, another blow for a region that until then had supplied almost half of the softwoods in the country. Yellow pine had been in trouble since the early 1970s, and air pollution was suspect. But before the cause could even be determined, the heat waves brought the forest to its knees, and the blistering summers that gripped the region after the turn of the century administered the coup de grace, turning forest into scrubland.

By 2015 more than 30 percent of southeastern farmland had been abandoned. The persistent heat steamed Memphis, Jacksonville, and New Orleans at over a hundred degrees for twenty days that summer. Southerners were on edge, civility went out the window. People shot one another over parking spaces. Crime and drug use soared. Production dropped as factories sent workers home early, or for the entire day, and then in a move that many parts of the nation soon adopted, a computer company in Tampa changed the workday to a worknight, with 9:00 P.M. to 3:00 A.M. and 3:00 A.M. to 9:00 A.M. shifts. The almost constant need for air conditioning tripled electric bills, and for many retirees on fixed incomes, it was either go broke, or try to sell the condo at a loss and go back north, where conditions were little better, or join the line for an entry permit to Canada. The Miami Dolphins moved to Calgary, and the Tampa Bay Buccaneers joined the Atlanta Braves in Edmonton.

Clouds blotted out the winter sun in Florida, and heavy rains fell. Sewage plants broke down, and mosquitoes and cockroaches bred with sultry abandon. So did alligators, at first. As predicted, the heat skewed the sex ratio for those that stayed behind, and a preponderance of males were born. Other more aggressive alligators moved north rapidly and recently reached the Potomac River. In January 2040, thirty inches of rain fell on the Tennessee River valley, and an avalanche of water thundered down on Chattanooga. The river crested at fifty-six feet and swamped the city, drowning 2,600 and causing $1 billion in damage.

In the Atlantic, the Caribbean, and the Gulf of Mexico, warming sea-surface temperatures intensified the strength of tropical storms. On September 10, 1997, the first of the super hurricanes, Hurricane Pierre, sent a storm surge of water thirty feet high and fifty miles wide smashing into the Florida Keys, drowning at least 22,000 people unable to evacuate on twenty-four–hour notice as Pierre suddenly changed course with wind speeds of 220 miles an hour. Intense hurricanes later devastated Vera Cruz, Corpus Christi, Galveston, Mobile, Fort Lauderdale, Hilton Head, Ocean City, eastern Long Island, Martha's Vineyard, and Nantucket. Typhoons raked the Indo-Pacific, killing an estimated 240,000 in the Philippines in 1996 and 700,000 in Bangladesh in 2005.

In the 1990s, slightly more than 3 billion people, half of the world's population, lived in the coastal areas of the Earth, and many of these areas were already overpopulated, polluted, and sinking into the sea. We now regard that time as almost benign. By 2030, the Mediterranean had risen a foot, and by 2050, two feet. Egypt has lost 15 percent of its farmland in the Nile Delta, and famine has become endemic. Bangladesh has surrendered 30 percent of its farmland to both advancing sea level and saltwater intrusion from the Bay of Bengal. The Indian Ocean laps ever higher over the Maldives, and in 2031 the government decided on complete abandonment of the islands, the first occasion in modern times that a population has had to flee its homeland because it was disappearing underwater.

The Netherlands is constantly reinforcing its sea defenses, and Britain has begun restructuring the Thames tidal barrage to protect London from flooding. Dutch engineers provided the expertise in the construction of the sea walls for Charleston and Boston, and they are also designing the tidal gates at the Narrows and Hell Gate in New York Harbor. The cost of defending cities against the ocean has already run into the tens of billions of dollars. As of 1990, the advancing sea level was eroding 80 percent of the East Coast beaches, most no more than ninety feet wide at high tide. As the years went by, each one-foot rise bit away at least a hundred-foot width of unprotected beach (and allowed storm surges farther inland). Financial resources were stretched to the limit, and certain areas had to be sacrificed. Massachusetts replenished Crane's Beach on the North Shore, and New York protected Jones Beach, while the Fire Island and Cape Cod national seashores were left to the mercies of a swelling Atlantic.

In the Chesapeake Bay, Long Island Sound, Narragansett Bay, and other coastal waters that once gave purpose and delight to sailors, swimmers, fishermen, and children building sand castles, the water is rankest in summer. A layer of sludge made of industrial wastes, sewage, blobs of oil, and silt blankets the inshore shallows. As the sun beats down and the slimy water warms, the sludge forms big gas-filled bubbles, "methane monsters," which rise to the surface and burst. Stimulated by warm, still weather, algal blooms spread for miles, then asphyxiate fish when the algae die off and deoxygenate the water below. Suddenly half a million menhaden, maybe a million, go belly up. After a week, the oily fish stink too much even for the gulls, but eels still burrow into the bloated fish to gorge on the internal organs. Above the rotting carcasses washed up on shore, next to the shattered houses half tumbled into the water, the air is filled with the ceaseless buzzing of greenhead flies and the scurrying of rats' feet over broken glass.

Apart from the heat waves that sent temperatures above ninety degrees for sixty-two days in Washington and forty-three days in New York City in 1995, summers in the Northeast were relatively cool during the decade. Starting in 2003, the muggy heat began in

May and didn't break until late October. (When Washington was hit with eighty-two days of ninety-plus temperatures in 2031, Congress finally voted to transfer the summer capital to Marquette, Michigan.)

The end of June 2004 kicked off an eight-day string of ninety-five–degree weather in the East. A dome of fetid air, which stretched from Louisville to Bangor, pressed down upon New York, a city where food prices escalated, water was rationed, and power outages were frequent. Smoke from dying forests as far as the Cumberland Plateau stung eyes and throats. With no beaches (except Jones, with its crowding and multiple stabbings and shootings), with the Palisades Interstate Park closed because of fire danger (it used to receive 7 million visitors a year, three times the number of Yellowstone), people were ready to pounce. The explosion came early in the afternoon on the Fourth of July. A band of homeless Iowa migrants camping in Washington Square burned an American flag under the arch, and a platoon of skinheads hired by developers to patrol lower Fifth Avenue attacked them with brass knuckles and tire irons. The fighting flared into a riot; when it stopped 212 people were dead and 1,400 injured. Other cities joined a move by New York to amend the Constitution by curbing freedom of assembly and restricting internal immigration, but their efforts failed.

More quietly, even larger problems were building up in the uninhabited parts of the world because the planet was warming fastest near the poles. Summer pack ice began retreating from Spitzbergen and the Siberian coast, and by 2007 shrinkage of the pack ice in the Arctic Sea generated a series of calamitous effects. First was the rapid disappearance or outright loss of marine and wildlife dependent on the food web based on the algae that grew on the underside of the pack ice. Populations of amphipods, mysid shrimp, and other zooplankton that had grazed on the algae collapsed, as did the schools of Arctic cod that had fed on them. In turn, the birds, foxes, seals, and narwhals that had relied on the cod died off. Walruses and polar bears suffered declines because they had used the sea ice as a bridge and as a platform to feed, hunt, and rest.

Development came to the North, and sleepy Churchill on Hudson Bay boomed, with the population soaring from 2,700 at the turn of the century to 340,000 now. Canada's most northerly deep-

sea port, Churchill offers the shortest shipping route from the heart-
land of North America to Europe. Starting in the mid-twentieth
century with completion of the railway from Regina, Churchill had
exported grain to Europe during the August–October navigation
season; but with the retreat of the pack ice and the expansion of
farmland and businesses in the prairie provinces, the port became
year-round, the busiest on the continent.

As in other Arctic boomtowns, explosive growth took a toll. The
large numbers of polar bears that used to gather in the Churchill
area during part of the year were extirpated. Without pack ice, the
bears could not move north, and, despite laws protecting them,
there was simply no way that they could coexist with the influx of
humans. In 2012, when a bear mauled and ate a shipyard welder,
dozens were shot within a month, and a mother and cub were set
ablaze with gasoline. After a bear died from eating a car battery at
a Churchill dump, people purposely poisoned their garbage to kill
more hungry bears. In 2017, the Canadian Fish and Wildlife Service
tranquilized seven of the beasts and airlifted them to Melville Island
near the magnetic pole; they gave the remaining four to zoos to be
bred in captivity. The last of them, Martha, died in the Cincinnati
Zoological Gardens on September 1, 2046, after an emergency generator failed during a week-long power blackout.

With pack ice no longer reflecting as much light from the mid-
night sun back into space, the Arctic Ocean heated quickly, causing
what scientists ironically termed a "positive" feedback. Similarly, the
growing warmth unexpectedly generated even more warming as it
sped the decay of tundra and taiga soils, which released more carbon dioxide and methane to the atmosphere.

The melting of the pack ice did not increase sea level because,
like ice in a glass, it displaced as much water as it contained. But
91 percent of glacial ice sits on Antarctica. Although disintegration
of the West Antarctic ice sheet will not occur before 2150, we can
do nothing to halt the process. By raising the sea level eighteen feet,
it will spell the end of coastal civilization.

The future looks yet grimmer. The feedbacks caused the National
Academy of Sciences to increase its warming projection to fifteen
degrees by 2090, but no one is taking this seriously anymore.
Humbled by decades of unpredicted catastrophes, scientists have

ruefully taken to recalling 1985, the year that Joe Farman surprised everyone by reporting a huge hole in the Antarctic ozone layer. No scientist had factored in the role that icy clouds play in abetting the depletion of ozone by man-made chemicals. The lesson then was, don't push the atmosphere into unexplored territory.

Wallace Broecker of the Lamont-Doherty Geological Observatory said it best in the late 1980s when speaking of the limitations of computer projections: "Earth's climate does not respond in a smooth and gradual way; rather it responds in sharp jumps. These jumps appear to involve large-scale reorganizations of the Earth system. If this reading of the natural record is correct, then we must consider the possibility that the major responses of the system to our greenhouse provocation will come in jumps whose timing and magnitude are unpredictable. Coping with this type of change is clearly a far more serious matter than coping with a gradual warming."

Instead of listening, politicians followed the advice of adaptationists who pointed to the gradual warming projected by a few scientists and said, "Why abandon fossil fuels? We can handle it." Now the will to act is finally here, but governments are too overwhelmed by today's catastrophes to plan for the future, a future that, for much of the world, may never come.

CHAPTER 2

Cause and Effect:

The Wages of Industrialization

Eᴀʀᴛʜ'ꜱ ᴄʟɪᴍᴀᴛᴇ ʜᴀꜱ ʙᴇᴇɴ ꜱᴛᴀʙʟᴇ ꜰᴏʀ ᴛᴇɴ ᴛʜᴏᴜꜱᴀɴᴅ ʏᴇᴀʀꜱ, ʏᴇᴛ even the small variations during this period, no more than a couple of degrees in either direction, have left a strong historical imprint on the world. The nineteenth century witnessed both the end of the last such fluctuation—the Little Ice Age—and the explosion in fossil-fuel use, which will entail much greater consequences.

In the past thousand years, the Earth underwent one additional climate change worth noting. During the so-called Medieval Warm Epoch between 800 and 1250, global mean temperature was about the same as it is now, but areas along the North Atlantic were a few degrees warmer, for reasons not fully understood. Vineyards flourished in England, and barley and oats were grown in Iceland. Sometime near the year 980, Eric the Red left Iceland and took advantage of ice-free northern waters to begin the Norse colonization of Greenland. About the year 1010, Eric's son, Leif, made the first of the Norse voyages to North America. These journeys ended about 1200 when the Norse settlements in western Greenland began to die out due to the return of frigid temperatures, and in 1492 Pope Alexander VI wrote that "because of the difficulty of passing through such immense quantities of ice . . . no vessel has touched there during the past 80 years." The Norwegian bishop whom the pope appointed never arrived, and John Davis, an English explorer who visited Greenland in 1585, found only Eskimos.

The world's climate gradually descended into the Little Ice Age, which lasted from about 1550 to 1850. The forward creep of glaciers and the thinning of tree rings indicate that temperatures dropped to their lowest point about 1700, roughly two degrees colder than today, then slowly but erratically began to recover. Sunspots may be a measure of the sun's intensity, and some scientists believe that their scarcity during this period means that the cause of the cold was reduced solar radiation.

British climatologist Hubert Lamb, who has studied manuscripts, diaries, logbooks, and paintings to document the effects of the Little Ice Age, notes that in 1563 Pieter Bruegel the Elder painted a benign nativity scene depicting the three kings offering gifts to a naked infant Jesus; four years later he painted a nativity scene in which peasants struggled through snow as the Holy Family huddled in the bitter cold. In sixteenth-century India, the monsoons became erratic, and lack of water prompted the abandonment of the great newly built city of Fatepur Sikri. Year-round snow, absent today, covered the high mountains of Ethiopia. The vineyards in England died out, the Thames froze over, Eskimos in kayaks reached the Orkneys—one even entered the river Don near Aberdeen. Recurrent famine in Scotland increased migration to North America and northern Ireland—by 1691, 100,000 Scots were in Ulster, planting the seed for the troubles that persist today.

Winters in colonial America were equally severe. Climatologist David M. Ludlum found that the winter of 1777–78, when American troops suffered at Valley Forge, was "notably mild" compared to others. But the winter of 1779–80 was called the worst "that was ever known by any person living," causing the Upper Bay in New York Harbor to freeze solid, which enabled the British on Manhattan to slide heavy cannons five miles across the ice to fortify Staten Island during the Revolutionary War.

As the Little Ice Age was receding, industrialization was getting under way. The global warming process was set in motion with the acceleration of the use of energy and a concurrent shift from wood to fossil fuels. The Chinese had burned coal long before the Industrial Revolution, but four centuries of fits and starts passed before coal burning took hold in England in the 1600s after the wide-spread clearing of forests. Other countries on the Continent gradually followed. In America, however, abundant forests provided both pioneer and city dweller with all the wood they needed for decades after others had turned to coal. As Richard G. Lillard wrote in *The Great Forest,* "All cabin dwellers gloried in the warmth of their fireplaces, exploiting their world of surplus trees where a poor man, even a plantation slave, could burn bigger fires than most noblemen in Europe. . . . The kind of hospitable settler who burned a whole log in order to boil a kettle of tea didn't consider his fire psychologically good until he had crammed a quarter of a cord into a space eight feet wide and four feet deep and had a small-scale forest fire roaring in front of him." This sort of excessive energy use still prevails in America today.

As late as 1870, wood supplied 75 percent of the energy used in the United States, but as forests were gradually stripped and the price of wood rose, Americans turned reluctantly to coal. Carbon-dioxide releases had grown at the beginning of the nineteenth century with the shift from water power to steam power generated by wood burning. The second great energy transition was even more decisive. By 1885, coal was king, fueling steamships, railroads, steel mills, factories, and homes, to power what Mark Twain called "the drive and push and rush and struggle of the raging, tearing, booming nineteenth century."

Rail lines expanded at an astounding rate (by 1870, the United States already had more than half of the track mileage in the world), and demand for electricity surged with a series of inventions, including Bell's telephone in 1876, Edison's electric light in 1879, and Nikola Tesla's electric motor, manufactured by George Westinghouse in 1888. In Edison's words, "the biggest and most responsible thing" he ever did was to build the world's first electrical generating station in 1882 on Pearl Street in downtown Manhattan, powered by coal. As Wilson Clark has noted, "The electric light and the electric power distribution system required to sustain it were to change the world as have no other technological developments."

By the end of the nineteenth century, more than 2,000 electric power stations were in operation in the United States, and every large city in North America, South America, and Europe had electric lighting. Electricity powered the newly built subways in Boston and New York City and propelled the trolleys into the suburbs and beyond. By 1910, a passenger, if so inclined, could travel from New York City to Portland, Maine, by a string of intercity trolleys.

In the atmosphere, carbon-dioxide levels had climbed nearly 7 percent, unnoticed. But the foundations were already being laid for another energy shift that would contribute to the decline of the trolley within a few decades and lead to yet another unprecedented spurt in emissions. Until the turn of the century, oil was the source of only about 3 percent of the energy used in the United States, but as electric lights began replacing kerosene lamps, the distillers looked for a new market, and what had been a useless petroleum by-product, gasoline, found a niche fueling the internal-combustion engine of the automobile.

The first car powered by gasoline was built in 1885, but steam and electric cars continued to dominate the market for another fifteen years. In 1895, after a gasoline car won the *Chicago Times-Herald* auto race, Henry Ford, a Michigan farmboy turned mechanic, became intrigued by its possibilities. He went to work for the Detroit Edison Company to learn more about electricity, became the best engineer in the plant, and was invited to attend the annual convention of Edison companies in New York. When Ford met his hero, the great Thomas Alva Edison himself, he sketched out his design of a gasoline engine on the back of a menu. "Young man, you have the right idea. Keep right at it," said Edison, slapping Ford on the back. "This car has an advantage over the electric car because it carries its own power with it."

Encouraged, Ford soon quit his job and sought backing by running his inventions in auto races, which were, as he wrote in his autobiography, "advertising of the only kind that people cared to read." In 1903, he built the eighty-horsepower "999," yet the tiller-steered racer was too much for Ford, himself a competitive driver, to handle. "Going over Niagara Falls would have been but a pastime after a ride," he wrote; and he hired Barney Oldfield, a professional bicyclist, to drive it in a race. After one week of lessons, Oldfield, who had never driven an automobile before, finished half a mile ahead of the nearest competitor. A week later Ford

21

founded his motor company, and shortly thereafter Oldfield became the first American to drive an automobile a mile a minute.

By 1913, mass-production techniques enabled Ford's employees to build a Model T in twelve and a half hours; but with the introduction of the assembly line the next year, a Tin Lizzie could be turned out every ninety-three minutes. Thanks to these and other innovations, Ford lowered the price of his models, and sales soared. Between 1910 and 1924 the price of the Model T dropped from $950 to $210, and Ford boasted, "Every time I reduce the charge for our car by one dollar, I get a thousand new buyers."

These developments can also be read in the histories of fuel consumption and carbon-dioxide buildup. During the first twenty years of the century, gasoline consumption in the United States jumped from 4,000 gallons to 4 million gallons. By 1923, there were more cars in Kansas than in either France or Germany, and by 1927 the United States was producing about 85 percent of the cars in the world. The age of the automobile had arrived, and, as Will Rogers said of Ford, "It will take a hundred years to tell whether he helped us or hurt us, but he certainly didn't leave us where he found us."

FATAL FLAWS

After World War II, oil supplanted coal as the world's dominant energy source. Today, oil supplies 38 percent of global energy needs, coal 30 percent, and natural gas 20 percent. Despite their dominance, fossil fuels—particularly coal and oil—are fatally flawed, and these flaws may be about to precipitate the next energy shift. For more than a century, human beings have largely tolerated malodorous, unhealthful, and destructive levels of air pollution from fuel combustion. Yet these may come to seem like relatively trivial problems when compared with the potential consequences of dramatically increased levels of carbon dioxide and other greenhouse gases.

Under natural conditions, light from the sun passes through the atmosphere to the Earth's surface, where it is absorbed and converted to heat. Some of this heat then escapes back into space in the form of invisible infrared radiation, without interference. If all of the heat were lost, the global temperature would be about zero degrees; but atmospheric water vapor, carbon dioxide, and a few other gases trap enough infrared radiation to maintain a stable mean temperature of fifty-nine degrees Fahrenheit. Now, with the total amount of carbon dioxide and other heat-trapping gases growing, the Earth's atmosphere may be compared to an overheated greenhouse.* The extreme case of an atmosphere saturated with carbon dioxide is Venus. The temperature at its surface is 840 degrees, hot enough to melt tin, lead, and zinc.

Carbon is essential to life and is present in the organic matter from which every living thing is made. When organic matter burns or oxidizes, its carbon combines with oxygen to form carbon dioxide, a colorless, odorless, nontoxic gas. Carbon dioxide, symbolized as CO_2, is made up of one carbon atom bound to two oxygen atoms. Beverage manufacturers use it to carbonate soft drinks; ice-cream vendors use it in its frozen form to refrigerate their products.

Green plants build their organic matter out of carbon dioxide absorbed from the air and water (H_2O) drawn from the ground. By taking energy from the sun during photosynthesis, they convert these molecules into oxygen, which they release, and carbon and hydrogen, which they store in their tissues in compounds such as carbohydrates and proteins. The sun's energy is stored in the chemical bonds of these substances at the same time. When these plants die and decay, most of their carbon oxidizes and quickly returns to the atmosphere in the form of carbon dioxide, but some organic matter is buried in soils and as sediments at the bottoms of lakes, swamps, coastal marshes, and shallow seas. Microbial action, heat, and the pressure of burial slowly "fossilize" this plant matter under anaerobic (air free) conditions, converting it into coal, oil, and nat-

---

*The greenhouse analogy is imperfect. Both the glass in a greenhouse and greenhouse gases like carbon dioxide allow the passage of sunlight through to the earth and then keep heat from escaping. But heat is carried upward from the earth in various forms. The glass in a greenhouse traps heat by confining warm air rising from the ground. On the other hand, greenhouse gases trap heat being carried off by infrared radiation. Nevertheless, the term "greenhouse effect" is used by scientists and is known to the public, so we retain it here. As the panes in a greenhouse are closed, less heat escapes and its temperature increases. Similarly, as the level of greenhouse gases rises, earth will warm up.

ural gas. (These fuels are also called hydrocarbons after their two main constituents, carbon and hydrogen.) Most of Earth's bituminous coal and a considerable portion of its oil originated in the abundance of plant life that existed during the Carboniferous period, 270 to 350 million years ago. When fossil fuels are burned in air, the photosynthetic process is reversed. The carbon and hydrogen originally captured by plants millions of years ago combine with oxygen and are released as carbon dioxide and water. About half of this excess carbon dioxide is soon absorbed by the ocean and the forests, while the other half remains in the air for centuries. As the chemical bonds in the organic matter break apart, the sun's energy is released, too, in the form of heat.*

The release of carbon dioxide via fossil-fuel burning can be viewed as an artificially accelerated version of a natural process whereby carbon moves from atmosphere to plants to soils and the sea and back to the atmosphere; this circulation is called the global carbon cycle. The seasonal increase in plant life in the spring, which pulls carbon out of the atmosphere, and the subsequent dieback in the fall, which releases it, are part of this cycle. On a longer time scale, carbon that was accumulated in organic matter and buried deeply in sediments over tens or hundreds of millions of years might return just as gradually to the air as carbon dioxide in the gas belched from volcanoes.

The difficulty is in this timetable, because human intervention through fossil-fuel combustion is transferring this sedimentary carbon into the atmosphere in just a few centuries. The rest of the cycle, which removes it very slowly, can't keep pace with this overload. If emissions were closer to natural levels, the oceans would absorb almost all of the excess. Instead, nearly 6 billion tons of carbon were emitted into the atmosphere in 1989 due to fossil-fuel burning, and emissions continue to grow by 2 percent per year, so the buildup of carbon dioxide is accelerating.

*Coal is largely carbon, so most of the energy released when it burns is accompanied by carbon dioxide. Oil contains more hydrogen, and when it is burned, much of the energy released is associated with the formation of water. Natural gas contains even more hydrogen, so even more of the energy released is associated with water. As a result, coal burning releases one-third more carbon dioxide for every unit of energy obtained than does oil and almost twice as much as natural gas.

There is no way to extract useful energy from fossil fuels without forming carbon dioxide and other harmful gases. For example, partially burned carbon is also released as carbon monoxide, CO, particularly from inefficient engines like those in automobiles. Coal and oil, but not natural gas, contain nitrogen drawn for nutrition into the living ancestral plants and sulfur incorporated during their decay. These substances are emitted as sulfur dioxide and nitrogen oxides, the progenitors of acid rain. Nitrogen oxides react in sunlight with carbon monoxide and various organic compounds left over from automobile combustion, to produce the brownish-orange mélange known as photochemical smog. Carbon monoxide, sulfur dioxide, nitrogen oxides, and organic compounds, while inherent in fossil-fuel burning, can be partially removed after combustion by scrubbers or catalytic converters. Carbon dioxide can be scrubbed too, but only at large sources like power plants—not in cars, for example—and only at a very high price.

The widespread destruction of forests has also added to the atmospheric build-up of carbon dioxide. Downed trees are no longer able to absorb the gas, and when burned their stores of carbon are released to the atmosphere as carbon dioxide. An acre of trees stores about 100 tons of carbon, and in the past forty years perhaps as much as half of the tropical forests on Earth have been destroyed. In 1987, slash-and-burn operations in Brazil covered an area almost the size of Maine. At the current rate of deforestation, most of the Amazon forest, still largely intact, may literally go up in smoke within the next fifty years. For every five tons of carbon emitted to the atmosphere by fossil fuels, at least one more arises from deforestation.

Even before humans began tinkering with the atmosphere, variations in the global carbon cycle produced natural fluctuations in the amount of carbon dioxide in the air. These occurred synchronously with small shifts in Earth's orbit that repeat every 20,000, 40,000, and 100,000 years. Orbital variations cause slight changes in the seasonal pattern of sunlight reaching the Earth. These subtle changes in sunlight apparently trigger large variations in atmospheric carbon dioxide, but how they do so is not precisely understood. One theory holds that more or less sunlight at certain places and times stimulates more or less photosynthesis by small marine plants called phytoplankton. The growth of these plants affects the carbon cycle by transforming carbon dioxide in the atmosphere into organic carbon in the ocean. When the plants die, much of the carbon sinks into the deep ocean rather than returning to the air, so an increase or decrease in photosynthesis would lead to a decrease or increase in the natural greenhouse effect.

We know how levels of carbon dioxide varied in the atmosphere by analyzing samples of ancient air buried deep in Antarctic ice (see the data provided in figure 2.1). When snow falls, air is trapped between the flakes, and when it is compacted into ice, this air is preserved in bubbles. Incredibly, the temperature of the clouds from which the snow fell can also be inferred from the particular composition of the ice. This record shows that Earth's temperature generally increased with the buildup of atmospheric carbon-dioxide concentrations and decreased with their decline, going back at least 160,000 years.

As it happens, the Earth is now in a warm cycle, which actually peaked about 6,500 years ago. The previous maximum occurred about 120,000 years ago. At neither time was Earth more than two or three degrees warmer than today. But overall, changes triggered by orbital shifts were much greater than those of the past thousand years. Between these two warm periods, Earth was gripped by a major ice age; it was as much as nine degrees cooler, while carbon-dioxide levels were as low as one-half the current amount. This parallel movement of carbon dioxide and temperature in the historical record supports the greenhouse theory, and it is reasonable to expect the two to march in lockstep in the future, as fossil-fuel emissions drive the level of carbon dioxide upward.

Variations in the level of carbon dioxide are accompanied by other changes, which alter its effect on climate. For example, water vapor is the most powerful of all greenhouse gases, due in part to its high concentration in the atmosphere. The level of water vapor is naturally thirty times higher than that of carbon dioxide, too high to be affected directly by human emissions. But warming caused by the addition of carbon dioxide leads to the evaporation of more water from the oceans and land to the atmosphere, creating a "positive feedback" by making the air more humid and trapping more heat. Another feedback, which may be either positive or negative, arises

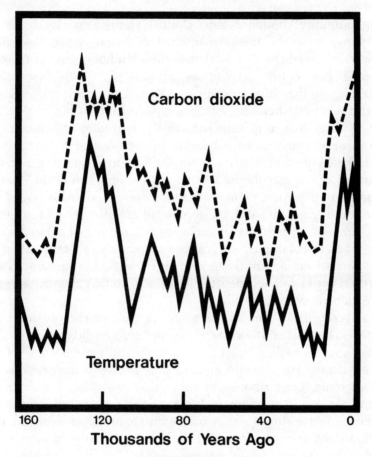

160   120   80   40   0
Thousands of Years Ago

FIGURE 2.1

Antarctic Ice-Core Record The close relationship between carbon dioxide and temperature is shown by these data from air bubbles trapped in Antarctic ice. As carbon dioxide varied by 110 parts per million, global temperatures ranged over about 11 degrees Fahrenheit. SOURCE: Adapted from J. M. Barnola, D. Raynaud, Y. S. Korotkevich, and C. Lorius, "Vostok Ice Core Provides 160,000-Year Record of Atmospheric CO_2," *Nature* 329 (October 1987): 408.

when humidity condenses into clouds. High, thin clouds act like greenhouse gases by trapping infrared radiation, while low, dense ones reflect sunlight, so a change in cloudiness brought on by warming can either amplify or reduce the greenhouse effect, depending on the altitude where it occurs. Yet a third feedback, also positive, occurs when the Arctic ice pack contracts in response to heating. This change in turn reduces the reflection of sunlight back into space and increases the heating of the ocean.

Feedbacks present a major complication in forecasting the consequences of the greenhouse effect, or interpreting in detail what has occurred in the past. Unlike the absorption of radiation by the greenhouse gases themselves, feedbacks are not well understood. They are not entirely accounted for in computer simulations, and they may act to amplify the expected warming, so they entail the possibility of rapid, surprising climate changes. By the same token, we could get lucky, and negative feedbacks could engender an unexpected slowdown of warming.

The really tricky part comes when the role of the oceans is considered. The oceans absorb some of the carbon dioxide emitted by human sources such as fossil-fuel burning, and natural ones, such as organic decay. The absorption of carbon dioxide is mediated by various currents, great ribbons of water that move like a rollercoaster around the world, sometimes breaking the surface, sometimes diving deep. Water surging from below brings nutrients from the deep ocean to the surface, fertilizing the phytoplankton blooms which regulate the absorption of carbon dioxide. But the currents themselves are sensitive to temperature, and an initial heating could be amplified if it altered their flow. For example, if global warming were to slow the upward movement of the rollercoaster, the surge of nutrients from the deep ocean would be reduced and so would the removal of carbon dioxide from the atmosphere by phytoplankton. Then the greenhouse effect would increase even faster.

This temperature-current linkage may cause climate to change rapidly and erratically. One theory proposes that as the Ice Age ended, the warming caused fresh water to melt from ice sheets rimming the North Atlantic, diluting a saltier, heavier current flowing from the equator that then heated the region before sinking to the bottom. The diluted water was less dense and couldn't sink, causing the downward motion of the rollercoaster to reverse. As a result, warm water couldn't enter the region, and the North Atlantic cooled quickly. But the cooling caused the ice sheets to stop melting, and the whole process quickly reversed itself.

This flip-flopping of climate may have been reinforced by changes in the oceanic absorption of carbon dioxide, which was first accelerated, then slowed, as the rollercoaster switched directions. Fossils of small marine creatures show evidence of rapid temperature changes at that time, possibly initiated by such a sequence of events. It could happen again as emissions move the Earth away from the stable climate that has persisted for 10,000 years.

MORE TROUBLE

The growth of carbon dioxide has been paralleled by a buildup of other atmospheric troublemakers that cause a host of difficulties in addition to global warming. Man-made chlorofluorocarbons (CFCs), which are used as refrigerants, solvents, and blowing agents for foam plastics, are extraordinarily potent greenhouse gases. One molecule of CFC-11 or CFC-12 can trap as much heat as 10,000 molecules of carbon dioxide, which will create lasting problems because CFCs survive for about a century after their release. Furthermore, these gases are responsible for the depletion of the stratospheric ozone layer, lying fifteen miles above the Earth.

Tropospheric Ozone

Ozone (O_3) is itself a greenhouse gas, but with a split personality. The protective layer of ozone in the stratosphere is formed by the natural action of ultraviolet sunlight on oxygen. The harmful tropospheric, or lower atmospheric, ozone, the same molecule but lying at lower altitude, is largely the product of sunlight acting on air pollution. The two are physically separated by the tropopause, a barrier of dead air at 30,000 feet that slows upward movement of pollution to a crawl. It can take two years for a gas to cross the barrier, so there is no way that tropospheric ozone can replenish the stratospheric ozone layer because it has a limited life* of only a few weeks (as opposed to CFCs, for instance, which live so long that they can dawdle across to the stratosphere).

Besides being a greenhouse gas, tropospheric ozone is toxic to plants and humans and is the key component of photochemical smog.† Smog levels are highest in areas with sunny days, heavy traffic, and temperature inversions that trap pollutants, such as Mexico City and the Los Angeles Basin. But ozone and its precursors can travel so far from their source that smog now covers much of the eastern United States and western Europe. Even in remote areas concentrations are growing at a rate of 1 percent a year. In Europe, levels appear to be twice those measured at the turn of the century.

The first harmful effects of photochemical smog were observed in Los Angeles in 1942, when a strange disease caused the demise of a multimillion dollar ornamental flower industry. Over the years, smog from the Basin has also helped to kill hundreds of thousands of ponderosa pines to the east, in the San Bernardino Mountains. The immediate cause of death may be infestation by bark beetles, but the trees have become vulnerable from stress by smog. A nasty

---

*The life, or lifetime, of a chemical in the atmosphere is determined by a variety of factors, including how fast it reacts with other chemicals, or is destroyed by sunlight, or is washed to the ground by rain, or is absorbed by the ocean.
†The term "smog" was coined in 1905 by Dr. H. A. Des Voeux of the Coal Smoke Abatement Society in Great Britain to describe the mixture of smoke and fog, the yellow fog of Sherlock Holmes stories, that enshrouded London during the colder months of the year in Victorian times. During a thick fog that lasted a week in December 1873, an estimated 700 people died from its direct effects, another 19 drowned after they mistakenly walked into the Thames or Regent's Canal, and many cattle in the city for an exhibition were said to have died from suffocation. Although Londoners deplored the persistent fog, a number of writers, notably Charles Dickens, Arthur Conan Doyle, and T. S. Eliot, occasionally took inspiration from it. So did artists, including J. M. W. Turner and James McNeill Whistler, while Claude Monet purposely visited London during the winter to paint his Thames series. The Great Smog of December 1952, which saturated London for five days and caused an estimated 4,000 deaths, eventually forced Parliament to pass the Clean Air Act of 1956. Even though coal smoke wasn't involved in the automobile-related ozone pollution of Los Angeles, the May 1955 *Scientific American* used the term "smog" to describe the foul air in that city, as did Lewis Mumford in *The City in History,* published in 1961. The term stuck.

smog that can blanket 200,000 square miles is the prime suspect for the forest decline afflicting the eastern United States.

The potential consequence of ozone pollution to human health is even more disturbing. At levels just above the federal air-quality standard, which was not even achieved in more than 101 metropolitan areas across the United States between 1986 and 1988, ozone causes chest pains and irritation of the nasal passage in experimental subjects during vigorous exercise. At higher levels, common in the Los Angeles area, the same effects occur with individuals at rest. Recent studies on animals suggest that continuous exposure to ozone levels *below* the federal standard may lead to permanent and serious lung damage, in a process similar to premature aging. Warmer temperatures in the future will drive smog levels upward, as will increasing ultraviolet levels if CFCs aren't eliminated. Persistent, intense smog could prove to be the last straw for the old, the sick, or the very young, already weakened by unbearable summer heat.

Nitrous Oxide

Nitrous oxide (N_2O) is another greenhouse gas that also leads to the destruction of stratospheric ozone. Its depleting effect is much smaller than that of the CFCs, but its level in the atmosphere is slowly growing. Following its introduction in the 1840s as a dental anesthetic, nitrous oxide became known as "laughing gas" because of the euphoria it induced in patients. Like carbon, nitrogen arises as part of a natural cycle. It is an essential nutrient for plants, yet they are unable to utilize it in the pure, molecular form (N_2) that occurs in the atmosphere. So nitrogen is captured from the air by bacteria in the soil and converted by them to ammonium (NH_4) and then nitrate (NO_3), both of which plants are able to digest. Nitrous oxide is formed as a by-product of the second part of this process.

In modern agriculture, humans bypass one or both of these steps by directly applying ammonium or nitrate fertilizer to soil. Unfortunately, bacteria act on artificial ammonium fertilizer as they do on the natural chemical, and the use of these fertilizers is contributing to the atmospheric buildup of nitrous oxide above natural amounts.

Over the past century, atmospheric concentrations have increased by 10 percent, and at present the level is growing at a rate of about 0.3 percent, or 5 million tons every year.

Methane

Nitrous oxide is a minor player in the natural nitrogen cycle. Most nitrogen is returned to the atmosphere as N_2 when another set of bacteria release it from nitrate. Similarly, methane (CH_4) provides a secondary route for the return of carbon in organic matter to the atmosphere. Under anaerobic conditions, as are found in the bottom of a swamp, methane forms instead of carbon dioxide, which is why natural gas, primarily composed of methane, is found with coal and oil. Methane is a double troublemaker as a greenhouse gas and as a precursor of tropospheric ozone. More than half of the methane now in the atmosphere comes from human activities with the result that, over the course of the past 150 years, the atmospheric concentration of methane has jumped more than 100 percent. It should be no surprise that methane leaks from coal mines, drilling operations, and pipelines. Indirectly, carbon monoxide from automobile engines and from forest fires also adds to the atmospheric concentration of methane because it destroys hydroxyl radicals, natural chemicals in the air that ordinarily keep methane in check.

In addition, humans are busy creating anaerobic environments, which encourage growth of the very bacteria that generate methane. Rice paddies, for example, which are spread across much of the world, are anaerobic. So are "sanitary" landfills stuffed with garbage. Still another source of methane is the stomachs of ruminant cattle. "A cow eats so much that is hard to digest that it has a symbiotic relationship with bacteria inside its gut," Ralph Cicerone of the University of California, Irvine, explains, "There are four compartments in a cow's stomach, and the one known as the rumen is ideal for a class of anaerobic bacteria and protozoa that can't live in the presence of oxygen. As they break down the cow's food, they produce methane. When a cow belches or is flatulent (farts, in plain English), the methane goes out fast. Animal experts have known about this for a hundred years. This is something we can measure in a controlled booth. About a third of the human race have similar microorganisms in their stomachs, but the amounts of methane produced are small compared to that of cattle, sheep, goats, horses, buffalo, water buffalo or elephants, and there are now 1.6 billion cattle in the world." Hard as it may be to imagine, bacteria in the guts of termites munching on the logs of felled trees have received passing attention as contributors to this process as well.

Yet George Woodwell, director of the Woods Hole Research Center in Massachusetts, is highly skeptical that the amount of methane emanating from these sources is sufficient to explain the observed atmospheric increase in the gas. Woodwell points out that bacteria in anaerobic soils work faster on dead organic matter at higher temperatures, creating more methane, and he concludes that the growth in methane has largely resulted from the recent warming trend. Indeed, evidence from the ice cores shows that the atmospheric concentration of methane has decreased when the Earth cooled and has increased when it warmed. If true, Woodwell's theory bodes ill for the future, for the initial warming from increasing carbon-dioxide levels could then spur the emission of even more methane. Moreover, another potential positive feedback of this sort may arise from the accelerated respiration of plants. Much as an overheated runner will draw on body fat for emergency calories, trees will consume their own carbon reserves faster as they warm, expelling more carbon dioxide into the atmosphere.

Consider the enormous human-induced interference with the atmosphere we have described: carbon-dioxide levels up by one-fourth; tropospheric ozone multiplied by two; methane by more; ozone depleted in the stratosphere; a totally artificial group of chemicals, CFCs, introduced and running rampant; a slowly rising temperature; and perhaps even the rate of plant respiration changed. At this juncture, anyone with a little common sense should be worried about the future of the Earth.

CHAPTER 3

The Learning Curve:

Scientific Discovery and the Political Response

ONE MIGHT EASILY ASSUME THAT HUMANS HAVE ONLY RECENTLY understood the negative consequences of polluting the atmosphere, which they otherwise would have averted. In fact, the case against fossil fuels has been building since 1285, when a commission was appointed in London to study the nuisance created by coal smoke. In 1661, John Evelyn wrote in his book *Fumifugium: The Inconveniencie of the Aer and the Smoak of London Dissipated* that human health and well-being as well as plants, buildings, monuments, and waters were being ruined by coal smoke.

Concern over global warming is also nothing new, and in fact the greenhouse effect has intrigued scientists since nearly the time of the Industrial Revolution. In 1827, the great French mathematician and physicist Baron Jean-Baptiste-Joseph Fourier set out the analogy between the behavior of heat in the atmosphere and its behavior in a greenhouse. The theory that rising concentrations of atmospheric carbon dioxide would increase this effect and lead to global warming was first advanced by Svante Arrhenius, a Swedish chemist, in 1896. He estimated that the temperature of the Earth would increase by four to six degrees Celsius (seven to ten degrees Fahrenheit) after the burning of fossil fuels had doubled the level of carbon dioxide in the atmosphere. Although Arrhenius later won the Nobel Prize in chemistry in 1903 for his theory of ionization, his prediction of global warming was largely ignored for over half a century until the change in carbon-dioxide levels was directly detected.

One of the few scientists to pursue the problem was Alfred J. Lotka, an American physicist who also developed one of the fundamental formulas of population biology, describing predator-prey relationships. In 1924 Lotka noted that the industrial era was dependent on exploiting fossil fuels accumulated in past geological ages, and he warned, "Economically we are living on our capital; biologically we are changing radically the complexion of our share in the carbon cycle by throwing into the atmosphere, from coal fires and metallurgical furnaces, ten times as much carbon dioxide as in the natural biological process of breathing." These human activities alone, he wrote, would in time "double the amount of carbon dioxide in the atmosphere," but in basing his calculations on coal use at 1920 rates, Lotka estimated that the doubling time was 500 years away.

G. D. Callendar, a British meteorologist who had gathered temperature records from more than 200 weather stations around the world, sought to persuade the Royal Society of London in 1938 that the global warming that had occurred since the 1880s was the result of increasing carbon-dioxide levels. His argument, however, was met with skepticism. Callendar went on to suggest that a 10-percent increase in atmospheric carbon dioxide would account for the observed warming, though there were at that time no ice-core measurements to establish the actual trend.

In 1954, G. Evelyn Hutchinson at Yale University suggested that the destruction of forests would also increase atmospheric carbon dioxide, but even scientists interested in the environment paid little attention. At the time, the Northern Hemisphere was cooling slightly, and ecological worries focused on other, more topical problems such as population growth, nuclear fallout, soil erosion, and water pollution. Thus in 1955, when an international symposium was held at Princeton on "Man's Role in Changing the Face of the Earth," not one of the participants, including some of the most distinguished scientists of the day, mentioned the greenhouse effect.

Perhaps this unconcern was also due to the general scientific agreement at the time that the ocean, which has a large capacity to store carbon, would harmlessly absorb almost all of the carbon dioxide emitted. But in 1957, Roger Revelle and Hans E. Suess of the Scripps Institute of Oceanography published a landmark paper on the carbon-dioxide exchange between the atmosphere and the ocean, in which they reported that the ocean had *not* absorbed as much carbon dioxide as previously assumed. Significant amounts would remain in the atmosphere and could eventually warm the Earth. They wrote that "human beings are now carrying out a large scale geophysical experiment of a kind that could not have happened in the past nor be reproduced in the future. Within a few centuries we are returning to the atmosphere and oceans the concentrated organic carbon stored in the sedimentary rocks over hundreds of millions of years. This experiment, if adequately documented, may yield a far-reaching insight into the processes determining weather and climate."

Shortly thereafter, C. David Keeling, a young colleague of Revelle's at Scripps, began to measure atmospheric carbon dioxide directly at the Mauna Loa Observatory, situated 11,000 feet high in Hawaii and far from any man-made sources of pollution. He found that the atmospheric concentration of carbon dioxide was at 315 parts per million (ppm), or about 0.03 percent of nitrogen and oxygen, and subsequent measurements disclosed that this concentration was increasing annually. As of 1988 it had reached 350 ppm, a 25-percent rise above the preindustrial level of 280 ppm that obtained around 1800, and which was only recently determined from the ice-core record (see figure 3.1). A doubling of the historical norm will occur by about 2075 if present trends continue.

Still, Revelle and Suess's findings didn't stimulate a sustained response, even among scientists. A Conservation Foundation conference held in 1963 on the "Implications of Rising Carbon Dioxide Content of the Atmosphere" suggested that "The effects of the continuing rise in atmospheric CO_2, while not now alarming, are likely to become so if the rise continues." But in 1965, the Conservation Foundation held another conference, this one on "Future Environments of North America." Not one of the participants discussed global warming, and there was merely a brief mention of the buildup of atmospheric carbon dioxide, which came when ecologist F. Fraser Darling, the foundation's vice-president, wondered if there

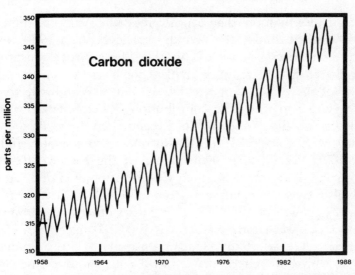

FIGURE 3.1
Carbon-Dioxide Levels for the Past Thirty Years Concentrations of atmospheric carbon dioxide measured since 1958 at Mauna Loa Observatory by C. D. Keeling and colleagues. The annual oscillations reflect the seasonal changes in Northern Hemisphere levels, which arise because plants store carbon by photosynthesis, primarily during summer, and release it back into the atmosphere at other times. SOURCE: C. D. Keeling, in *Policy Options for Stabilizing Global Climate,* U.S. Environmental Protection Agency, February 1989, p. 3.

had been "some weather modifications" because of "the increasing carbon-dioxide content of the air." Paul E. Waggoner, a meteorologist and plant pathologist at the Connecticut Agricultural Experiment Station, ended any further discussion by stating, "So far the increase in carbon dioxide with time in the open country is still so small that there are people who don't believe there has been one. This is reassuring."

But with the accumulation of Keeling's rising CO_2 measurements and other findings, scientific interest began, slowly, to grow. In 1967 Syukuro Manabe and Richard Wetherald of the Geophysical Fluid Dynamics Laboratory in Princeton, New Jersey, used a computer simulation to calculate that the average global temperature might increase by more than four degrees when atmospheric carbon dioxide reached double the preindustrial level. At a 1970 environmental conference sponsored by the Massachusetts Institute of Technology, global warming ranked high as a concern, partly because of the presence of Keeling, as well as Woodwell, who was just beginning to study the contribution of deforestation to atmospheric carbon dioxide levels. In Sweden, Bert Bolin of the International Meteorological Institute was undertaking similar studies.

In 1975, Veerabhadran Ramanathan at the National Center for Atmospheric Research in Boulder, Colorado, pointed out that CFCs could also cause a significant greenhouse effect. The next year, before the growth of infrared-trapping gases other than carbon dioxide and CFCs was clearly established, W. C. Wang, James Hansen, and their colleagues at the Goddard Institute of Space Studies and Harvard University argued that a broad range of trace gases, particularly nitrous oxide, methane, and CFCs, might increase as a result of human activities and that their combined effect on climate could become comparable to that of carbon dioxide. By 1981, after the levels of trace gases had been tracked for several years and were clearly increasing, Hansen and his colleagues demonstrated that in concert, CFCs, methane, and others were already as important as carbon dioxide. They wrote most prophetically, "The combined warming of carbon dioxide and trace gases should exceed natural temperature variability in the 1980s and cause the global mean temperature to rise above the maximum of the late 1930s."* The outspoken Hansen has his critics, but he and his colleagues were right on the mark when they made this judgment.

U.S. government interest in global warming can be traced to 1979 when, after the fall of the shah of Iran, oil prices rose and inflation hit double digits. In response to the public clamor over the long lines at gasoline stations, there was agitation in Congress for the development of synthetic fuels. Oil and gas could be manufactured

---

*To appreciate the above prediction, recall that the 1940–1970 hiatus in warming had barely ended and the temperature surge of the 1980s had not yet begun. Recent calculations attribute 60, 20, 15, and 5 percent of the current manmade greenhouse effect to carbon dioxide, methane, CFCs, and nitrous oxide, respectively. The future ratios will depend on the course of actual emissions from here onward.

from domestic coal and oil shale, which would assure the United States of adequate supplies. President Jimmy Carter reacted by proposing a massive $88 billion program to be administered by the federally supported United States Synthetic Fuels Corporation (or Synfuels Corporation).

Shortly afterward, Revelle, Keeling, Woodwell, and geophysicist Gordon MacDonald submitted a warning to the White House that the heat required to make synthetic fuels would itself require fuel combustion, and thus would double the carbon-dioxide emissions from using oil. They cautioned that "man is setting in motion a series of events that seem certain to cause a significant warming of world climates unless mitigating steps are taken immediately."

At the request of the White House, the National Academy of Sciences set up a review panel on the greenhouse effect, which advised that "a wait-and-see attitude may mean waiting until it's too late" to avoid major climate changes. An amendment attached to the Energy Security Act of 1980 established the Synfuels Corporation, mandating that the National Academy conduct a complete study of the greenhouse problem. Twenty-five years of scientific investigation since the Revelle and Seuss paper was finally causing a minor ripple in Washington's policy circles.

The Academy report came out in October 1983. It confirmed the finding of the earlier assessment that a doubling of carbon-dioxide levels eventually would warm the Earth by three to eight degrees. Nevertheless, no available evidence "would support steps to alter current energy use away from fossil fuels." "Overall," the authors wrote, "we find the CO_2 issue reason for concern, but not panic," for "both climate change and increased CO_2 may also bring benefits." The panel included Roger Revelle and four other marine or atmospheric scientists, but it appears to have been particularly influenced by Paul Waggoner and two economists, William Nordhaus of Yale and Thomas Schelling of Harvard. Drowned out was the troubled voice of ecologist Woodwell, who also served on the panel. But the Academy report was released a shade too late to preempt a debate. The Environmental Protection Agency's little-known strategic studies staff had months before prepared a study called *Can We Delay A Greenhouse Warming?* which concluded that by 2030 all of the greenhouse gases together would create the same effect as doubling carbon dioxide alone, though the resulting warming would be delayed for a few decades because the oceans would be slow to heat (a point we explore in chapter 5). "As a result," the EPA study held, "agricultural conditions will be significantly altered, environmental and economic systems potentially disrupted, and political institutions stressed." The EPA study hit the newspapers the day before the Academy report.

With the Academy report urging a relaxed, academic approach and the EPA study forcefully arguing that it was too late to avoid big shifts and that planning for adaptation to a changed world should begin, a public debate on global warming was finally joined. Concern had been intensifying elsewhere as well. Beginning in 1980, three international agencies, including the United Nations Environment Programme, sponsored a series of conferences and assessments, which culminated at a meeting in Villach, Austria, in the fall of 1985. With leadership from Bert Bolin of Sweden, the scientific community was finally willing to present a clear consensus position to the public. "Many important economic and social decisions are being made today on long-term projects . . . such as irrigation and hydro-power; drought relief; agricultural land use; structural designs and coastal engineering projects; and energy planning—all based on the assumption that past climatic data . . . are a reliable guide to the future. This is no longer a good assumption." The conference report warned that "while some warming of climate now appears inevitable due to past actions, the rate and degree of future warming could be profoundly affected by governmental policies on energy conservation, use of fossil fuels, and the emission of some (non-CO_2) greenhouse gases." It went on to recommend consideration of a global treaty to deal with climatic change.

THE WARNING SIGNAL

A revolution in attitudes about the atmosphere began on May 16, 1985, when Joe Farman and colleagues at the British Antarctic Survey published a paper in *Nature* reporting an enormous seasonal decline in the stratospheric ozone layer over Antarctica. The authors suggested that the decline was due to the presence of chlorofluoro-carbons which were known to have spread worldwide, while some other scientists believed that the peculiar natural motions of the atmosphere over Antarctica, the coldest place on Earth, were solely to blame.

The implications of ozone depletion on a large scale are cataclys-mic. As the primary filter of ultraviolet light, the ozone layer blocks excessive radiation from reaching the Earth, where it would other-wise ignite a pandemic of skin cancer, as well as a total collapse of the natural world. Life simply could not exist on Earth without the ozone layer. Farman had first noticed the hole in the ozone layer after studying 1982 data, but further analysis showed that a marked depletion had begun about 1977. From that point on, ozone levels had decreased each Antarctic spring, during September and Octo-ber, only to recover by the Austral summer; and ominously, the hole was growing larger over time.

Farman's measurements were made at Halley Bay in Antarctica with a spectrophotometer, a device that detects incoming levels of ultraviolet light that manage to penetrate the ozone layer. The amount detected will vary with the thickness of the layer. Cau-tiously, he had watched the hole enlarge for a couple of years and had even cross-checked his measurements by installing a new instru-ment before publishing his findings. Fortunately, while Farman's instrument was pointed upward toward the ozone layer from the ground, a NASA satellite was looking downward and measuring the ultraviolet light reflected backward into space, a process that also depends on the amount of ozone present. At first it appeared that nothing unusual had been detected, because NASA had pro-grammed its computers to ignore low ozone values on the assump-tion that they could not occur. Farman's findings prompted NASA scientists to take another look. Indeed, by that summer they had located the hole, and they confirmed Farman's observations. Sub-sequent measurements with the NASA instrument showed that at its lowest point only half the normal amount of ozone was present above Antarctica and that the area of depletion covered a region roughly the size of the United States. Blobs of low-ozone air break-ing free of the hole appeared to be causing significant depletion as far north as Tierra del Fuego, Patagonia, and southern Australia. It began to look as though the ozone hole could turn into a global threat.

CFCs

Chlorofluorocarbons had been invented by the chemist Thomas Midgley, who left another dubious legacy, lead additives for gasoline. One of the most honored scientists of his time, Midgley won medals for both discoveries. In 1923, the American Chemical Society, of which he later became president and chairman, gave him the William H. Nichols Medal for inventing leaded gasoline, and in 1937 the society joined with other scientific organizations in awarding him the Perkin Medal, the greatest honor an American chemist could receive aside from a Nobel, for discovering CFCs.

Midgley came up with CFCs after the Frigidaire division of General Motors asked him to find a safe replacement for the potentially toxic chemicals, sulfur dioxide and ammonia, then used to cool refrigerators. It took him and two assistants only three days to concoct the first CFC, a water-clear liquid when kept at low temperature, chemically known as dichlorodifluoromethane (CFC-12). Midgley and his assistants knew that it was harmless because on the last day, as reported by *Fortune* magazine, "They put a teaspoonful, furiously boiling at room temperature, under a bell jar with a guinea pig, while a physician watched earnestly for symptoms of the guinea pig's collapse. There were none."

Midgley announced the discovery at the 1930 convention of the American Chemical Society. A big, bespectacled, jovial man with a flair for the dramatic, he later demonstrated the nonexplosiveness of CFCs by inhaling a lungful of the warmed gas, holding his breath, and then blowing it through a rubber hose attached to a beaker with a lighted candle inside. The flame wavered and went out. An invalid from polio the last four years of his life, Midgley died in 1944 of accidental strangulation, after he became tangled in a harness he had devised to help lift himself from bed.

Placed on the market, CFCs were quickly hailed as a miracle compound (as were polychlorinated biphenyls—PCBs—marketed by Monsanto as dielectric fluid in that age of environmental naïveté). Besides serving as refrigerants, CFCs gradually found use as a gas to blow foam used in pillows, dashboards, or insulation; as propellants in aerosol spray cans; and as solvents to clean computer chips. Now CFCs are literally all over the place. Indeed, as part of its argument for the continued use of CFCs, a major U.S. industry trade group, the Alliance for Responsible CFC Policy, implied that national security was at risk because "the Pentagon would be uninhabitable without air conditioning during warmer months."

The versatility of CFCs originates in the two properties Midgley was so proud of, nonflammability and nontoxicity. But these two properties are also their downfall, because they mean that CFCs are so resistant to reaction with other chemicals that they are impervious to the natural processes that usually cleanse the atmosphere of pollutants. So when a discarded air conditioner or refrigerator is crushed by a bulldozer in the town dump, the CFCs are released into the air, where they can survive for decades while gradually infiltrating the stratosphere.

In 1974, F. Sherwood ("Sherry") Rowland and Mario Molina of the University of California, Irvine, were the first to warn about the consequences of CFCs ascending to the stratosphere. The chairman of the university's chemistry department, Rowland had an international reputation based on his studies of the chemistry of radioactive materials. But when he attended an Atomic Energy Commission meeting in Fort Lauderdale in 1972, he was feeling restless and was looking for new fields to explore. During a coffee break he learned from another conferee that James Lovelock, the independent and unorthodox British scientist best known today as the father of the Gaia hypothesis,* was about to report in *Nature* that he had measured CFCs in the lower atmosphere over both the Northern and the Southern Hemispheres.

Lovelock's interest in CFCs was largely academic. He had stumbled upon them while trying out an electron capture detector, a state-of-the-art measuring device that he himself had invented earlier. The only practical question he raised was whether CFCs survived in a glob of moving air long enough to track it, for he thought

---

*The Gaia hypothesis states that the atmosphere, the oceans, the climate, and the surface of the earth are regulated by the behavior of living organisms to maintain a state comfortable for life.

meteorologists might use the chemicals to trace the direction of winds. Establishing a monitoring station at home, Lovelock made his family stop using aerosol sprays so as not to interfere with his backyard CFC measurements. He also applied to a British government agency for a small grant to study CFCs from a ship sailing to South America. Although his application was rejected at first, some friendly civil servants dipped into a discretionary fund to pay his travel and subsistence expenses.

Lovelock was excited simply to find CFCs in the atmosphere over the South Atlantic; and in the article reporting his results, which was published in January 1973, he pronounced them to be "no conceivable hazard" and abandoned the field for other research. As Lovelock later admitted, "I boobed. It turned out I was sitting on a real bomb."

Rowland wondered about the eventual fate of all of those CFCs found by Lovelock. He and Molina, a postdoctoral research associate who had just received his Ph.D. from Berkeley, began their investigation in October 1973. At the time, the annual U.S. production of CFCs was on the order of 850 million pounds, and the major manufacturer was Du Pont, which sold them under the trade name Freon. By December, Rowland and Molina had completed their calculations, and in June 1974 their paper was published in *Nature*. The results of their research were startling, but as Rowland recalled afterward, "There was no moment when I yelled 'Eureka!' I just came home one night and told my wife, 'The work is going very well, but it looks like the end of the world.' "

Briefly put, Molina and Rowland noted that CFCs are being added to the environment in steadily increasing amounts, and, as they are not destroyed by any processes in the troposphere, they are able to survive for years while slowly drifting up into the stratosphere, ten to thirty miles above the Earth. Once there, CFCs are gradually decomposed by ultraviolet radiation that never reaches the lower atmosphere. Chlorine atoms are released in the process. The chlorine in turn triggers a catalytic chain reaction, in which a single chlorine atom can be used over and over again to destroy hundreds of thousands of ozone molecules before eventually falling back to Earth as hydrochloric acid and disappearing harmlessly into the sea.

The atmosphere is thicker near the ground, thinner above,

because the tug of gravity pulls the gases downward. So any molecule in air spends most of its life in the lower atmosphere, which is the region in which CFCs are not decomposed. Their passage through the higher altitudes, where decomposition occurs, is so infrequent that some CFCs live for more than 100 years.* As Rowland later wrote, "A 120-year average lifetime, without any intervening major changes in the atmosphere, means that 90 percent of the molecules now in the atmosphere will still be there by 2000 A.D., 39 percent by 2100 A.D., 17 percent by 2200 A.D. and 7 percent by 2300 A.D. Even without any further emission of [CFCs]—and releases are occurring daily all over the world sufficient to average about 400 kilotons annually—appreciable concentrations of [CFCs] will survive in the atmosphere for the next several centuries."

The Rowland–Molina paper prompted considerable controversy. A public boycott of aerosol spray cans, considered to be an unnecessary use of CFCs, spread almost immediately, and emissions began to decline. Robert Abplanalp, the biggest aerosol-spray manufacturer in the world, publicly lambasted the scientists but privately saw to it that his Precision Valve Corporation, in his words, "Got out in front in R and D" in the search for a substitute propellant.† Further research sponsored by both the federal government and industry confirmed the general outlines of the depletion theory, and by 1978, with some simple alternatives for the CFC propellant already available and others in the offing, four countries—the United States, Canada, Norway, and Sweden—had banned most applications of CFC-containing aerosol sprays. The ban was considered to be a prudent though modest response to a distant threat. Like the greenhouse effect, significant depletion of the ozone layer was thought to be decades away.

---

*Molecules in the air circulate like people seated in a ferris wheel. Suppose a madman has climbed a tower next to the wheel and is taking potshots at it. But any given person is in his sights only when he or she is at the top position of the wheel, and the madman's aim is poor to boot. Thus a person (or CFC molecule) might circulate many times before being hit by a bullet (or ultraviolet radiation).

†After the 1978 ban on aerosols went into effect, Abplanalp was ready with Aquasol, a nonflammable mixture of water and butane. "This gave us a hell of an edge," Abplanalp says. "Maybe I did call Molina an asshole, but he did a lot for us. Business has been fantastic."

Although CFCs then receded from the news, scientific interest in them did not end. In 1980, five years before Joe Farman published his findings, the United Nations Environment Programme convened a working group of experts to assess the CFC problem, and parallel political discussions on a possible global agreement on CFCs started in 1982. The United States pressed for a worldwide aerosol ban, but the European community resisted. Unable to agree on specific limits, the United States and twenty other countries signed the Vienna Convention for the Protection of the Ozone Layer in March 1985. This provided for the adoption of protocols at a later date which would limit or prevent the emission of ozone-depleting substances, should the scientific evaluations support such action.

More scientific workshops were convened. The EPA contributed an analysis that indicated that an immediate 85-percent cut in CFC emissions would be needed to stabilize the ozone layer, and that not even a complete ban on all CFC uses would ensure that the layer would be restored within a century or more. During one session early in 1986, Du Pont admitted that substitute chemicals for CFCs as refrigerants and in other applications could be manufactured if the company were given sufficient incentive to do so. Meanwhile, despite the aerosol ban, expansion of these other uses was causing CFC emissions to climb once again, by as much as 3 percent per year.

Following its confirmation of Joe Farman's discovery, NASA began displaying full-color renditions of the ozone hole that fascinated and frightened the public. And then the entire gestalt shifted. No longer was a minor depletion, still years down the road, at issue. The ozone layer was disappearing fast, and the fate of the planet was at stake. Scientists, however, still divided into two camps, one adhering to the theory that the natural dynamics of the atmosphere over Antarctica were the only cause of the hole, the other arguing that the low temperature and unusual air motions would create optimum conditions for chlorine reactions to proceed rapidly, but that ozone depletion simply would not occur unless CFCs were present.

In October 1986, a scientific expedition reported new ground-based observations from Antarctica by satellite link to a Washington news conference. The evidence wasn't absolute, but their findings conflicted with all proposed causes for the ozone loss, except reactions with chlorine from CFCs. Suspicions about CFCs increased in the following months as the data were analyzed again and again. Naturally occurring molecules of nitrate, which usually provide some protection for ozone by scouring the atmosphere of chlorine, were at unusually low concentrations. Scientists speculated that they had simply frozen out of the supercold air onto high-altitude stratospheric clouds that occur largely in polar regions, allowing chlorine to do its damage with unusual efficiency.

Policy makers began hearing a more concerned tone in scientists' voices, and Congress was growing restive. EPA Administrator Lee Thomas decided to urge a total phaseout of CFCs. After discussions between the EPA and the State Department, the United States took the position at the December opening session of protocol negotiations in Geneva that the industrialized nations of the world should cut CFC production by 95 percent within a decade or so.* A minor war ensued within the Reagan Administration. According to attorney David Doniger of the Natural Resources Defense Council, whose lawsuit helped keep the EPA on track toward developing CFC regulations, a swarm of midlevel White House officials, most notably from the Office of Management and Budget, vehemently opposed any such actions. Doniger noted that one OMB official suggested that skin cancer was a "voluntary" consequence of sun exposure, which might reasonably be ignored by the government. This line of thinking may have given rise to Interior Secretary Donald Hodel's purported remark that life-style changes (like hats and sunglasses) provided a possible alternative to regulation.

But an international network of environmentalists dispatched observers to the protocol negotiations and kept the spotlight on them for the public back home. With the State Department on his

---

*Thomas told MO that, even leaving the ozone hole aside, there was no reason not to go for a phaseout since production of substitutes for CFCs appeared to be feasible. His primary objective for the Geneva talks was negotiation of a schedule of reductions in CFCs which would convert the substitutes from feasible to commercially available. Thomas argues that while the ozone hole was becoming a progressively larger factor in his thinking during this period, its major role may have been in awakening the public here and abroad and causing them to pressure their political leaders into supporting a serious agreement.

side as well, Thomas eventually won the internecine warfare during the summer of 1987. European governments initially demurred at a full attack on CFCs, but a compromise became possible when the West Germans, under pressure from a burgeoning Green vote, broke ranks. In September of that year, the United States and twenty-three other nations formally signed an accord in Montreal calling for a 50-percent reduction in CFC production in the world's industrialized countries by mid-1999.*

The debate about the origin of the ozone hole was being resolved just as the ink was drying on the Montreal Protocol. James Anderson and his colleagues at Harvard had designed an elegant experiment to detect chlorine monoxide, the product of the reaction of ozone with chlorine that destroys ozone. Anderson's experiment was mounted aboard a modified U-2 airplane, which could operate somewhat like a hurricane spotter, flying into the ozone hole to gather information directly. The measurements showed that the ozone levels plunged (and, simultaneously, levels of chlorine monoxide soared) as the airplane entered the hole. His findings ended the dispute: it was the chlorine from CFCs that was destroying ozone (see figure 3.2).

More bad news was to come. In March 1988, the NASA Ozone Trends Panel first reported that ozone levels had declined 3 percent globally in twenty years, not just over Antarctica. The blobs of low-ozone air leaking from the hole were the suspected cause. Finally, in February 1989, an expedition to the Arctic reported that CFCs were creating a situation much like the ozone hole over that region, which is perilously near northern Europe. Robert Watson of NASA, who organized the Arctic expedition, bluntly called the finding "a strong message to policy makers."

Led by British Prime Minister Margaret Thatcher, the ministers from the European Community finally called for an end to CFCs,

---

*The reduction applies to CFC-11 and CFC-12, as well as to three other CFCs that have much lower levels in the atmosphere. Emissions of related chemicals, called halons, which contain bromine and are used in fire extinguishers, were frozen by the protocol, not reduced. Both the ozone depletion and the greenhouse effect caused by some related but uncontrolled chemicals could become substantial if they are used as substitutes for CFCs, so the protocol will need to be strengthened and the number of substances to which it applies broadened.

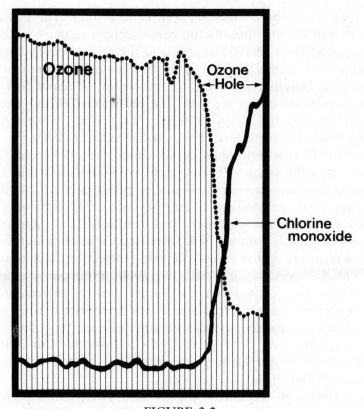

FIGURE 3.2

Anderson's Measurements in the Ozone Hole As the airplane carrying the measuring devices flew south at a fixed altitude toward Antarctica (left to right on this graph), ozone levels (shaded area) suddenly fell, marking the edge of the ozone hole. Simultaneously, chlorine monoxide levels rose inside the hole. SOURCE: Adapted from J. G. Anderson, W. H. Brune, and M. H. Proffitt, "Ozone Destruction by Chlorine Radicals Within the Antarctic Vortex: The Spatial and Temporal Evolution of ClO–O$_3$ Anticorrelation Based on in Situ ER-2 Data," *Journal of Geophysical Research* 94 (D9 [August 1989]): 11, 477.

and President George Bush followed close behind. The politicians were responding not just to the ozone depletion but to the blistering hot summer of 1988 as well, and the environmental concerns it had aroused in the public at large. In May 1989, eighty countries meeting in Helsinki, including the previously obdurate India and China, agreed informally to phase out the major CFCs by the year 2000.

The discovery that human beings could radically alter a basic feature of the Earth's atmosphere on which life depends was to change forever the public's attitude toward the environment. It had a profound impact on scientists as well. For many of them, what had previously been an intellectual exercise took on the character of a nightmare. About the same time, their confidence in computer projections of global warming was increasing, a joint Soviet–French team was reporting the carbon dioxide–temperature link found in the Antarctic ice-core data and the 1980s produced a string of record-hot years. John Firor, the former director of the National Center for Atmospheric Research, recalls how an atmospheric scientist who was studying global warming came into his office about this time, closed the door, carefully looked around, and confided, "I'm starting to believe the predictions of my model, and it's scaring the hell out of me."

Compared with world response to other international problems, such as terrorism, drugs, or nuclear-weapons proliferation, the response to the threat to the ozone layer was unusually quick, once direct evidence of damage was apparent. Does this mean that the nations of the world are now prepared to act swiftly to halt the catastrophic consequences of global warming? Not necessarily. A major obstacle impedes the path toward a political solution: persistent scientific uncertainty.

CHAPTER 4

Judges:

A Critique of Pure Fact

NINETEEN EIGHTY-EIGHT WITNESSED AN ATTEMPT TO END ANY doubt among politicians about the imminence of global warming. Testifying before the Senate Committee on Energy and Natural Resources, James Hansen of the Goddard Institute for Space Studies said bluntly, "It is time to stop waffling so much and say that the evidence is pretty strong that the greenhouse effect is here."

Hansen is an unassuming type, one who stops and thinks before answering a question, so his public candor, unusual for a scientist in any event, was all the more striking. It raised the hackles of several other climatologists, who thought he had stretched the facts too far. *Barron's* magazine quoted Reid Bryson of the University of Wisconsin as calling Hansen's testimony a "phenomenal snow job." Actually, what Hansen had to say was an exercise of sound scientific judgment.

Hansen was not the first scientist to decide that the 100-year warming trend could be ascribed to the greenhouse effect, but he was the first to say so in simple English, in a very public forum. Depending on how much confidence scientists place in their judgment, they may express themselves quietly while waiting for an elevator at a conference, or explicitly in a journal article that thirty people may ultimately read, or to a larger audience, if they have

51

access to one. If Hansen had said exactly the same thing in a scientific paper, the public at large might never have become aware of his conclusion, unless some astute reporter spotted it. In dealing with global warming, judgments like his, rendered in clear language, will be more than useful to the general public. They will be critical.

To understand this point, consider the question that Hansen put to himself in his testimony: Is the measured one-degree, 100-year warming of the Earth attributable to the increase in the atmospheric greenhouse gases? In order to answer this question, a scientist would proceed as follows. Several factors are known to influence or "force" climate over a 100-year period, including the eruption of volcanoes, variations in solar radiation, and emissions of greenhouse gases. If you enter the 1880s values of these factors into a computer simulation, or model,* of climate behavior and press the button, out pops an estimate of the 1880s global mean temperature. Now enter the 1980s values for all factors except the greenhouse gases; leave them the same as the 1880s. Press the button, and out pops an estimate of how much these other factors affected the temperature by the 1980s. Say the result is much smaller than one degree. Finally, enter the 1980s values for *all* factors, press the button, and out pops an estimate of the 1980s temperature. The question is, does the added effect of the increase in greenhouse-gas levels warm the Earth by one degree, or not? Unfortunately, things aren't that simple, for two reasons.

First, the number of stations around the world with acceptable temperature measurements has changed over time, so the reliability of the global mean temperature, which results from averaging these measurements, varies considerably. Hansen averages temperatures from as many as 1,800 continental and island stations to calculate the global mean during the 1960s, but uses as few as 300 for the 1890 value. In addition, measurement techniques have changed, weather stations have moved, and, in some cases, the surrounding environment has altered. For instance, the heat trapped by growing cities, which keeps them warmer than the surrounding countryside at night, has increasingly affected some readings. Temperatures of the sea surface, sometimes used in these analyses, were once deter-

---

*A model is a set of equations, generally solved by computer.

mined from a bucket of water hauled up on deck; today, they are measured as water is sucked in to cool a ship's engine. So today's temperature readings are not strictly comparable to the older values. Consequently, the measured change in global mean temperature is not a single quantity, but has a range of uncertainty.

Reasonable people disagree over the preferred value, because there are different approaches to evaluating the various confounding influences. Hansen's Goddard group cites a most likely value of 1.1 degrees Fahrenheit for the warming since the 1880s but acknowledges that the temperature change could lie anywhere between 1.5 and 0.8 degrees Fahrenheit. Tom Wigley's Climate Research Unit at the University of East Anglia in Great Britain analyzes both land and sea temperatures, which indicate a warming of about one degree Fahrenheit, slightly smaller than Hansen's reckoning but quite close to it, given the uncertainties.

Second, the computer models, which determine the climatic consequences of changes in the various forcing factors, are a hodgepodge of precise physical understanding, crude guesses, and calculator shortcuts. For example, feedback factors are only roughly understood (as discussed in chapter 2), and each model treats them a little differently. The mass of water in the ocean is slow to heat, retarding the warming, and the various models also account for this effect in different ways. As a result, projections of future warming range from 20 percent higher to 50 percent lower than the Goddard simulation followed in this book.

Say you agree with Hansen that the actual measured change was somewhere between 0.8 and 1.5 degrees. Say the computer could spit out a temperature change attributable to greenhouse gases of anywhere between 0.9 and 1.4 degrees, depending on how the crude guess factors in the model are manipulated. The agreement between your temperature estimate and the computer's calculations could look very good or very bad, depending on which crude guesses you prefer to use in the model and which reckoning of measurement approaches you prefer within the above range. It's not like flipping a coin or comparing 1.696 to 1.695 and deciding whether they are identical. As Stephen Schneider of the National Center for Atmospheric Research has noted, "Different scientists will retain or shed their skepticism [that the trend is related to the greenhouse effect] in different degrees and at different times." In effect, it's a matter of judgment.

There is one way around this quandary, one way to improve the quality of our judgment. Ground-level global temperature is not the only factor that has been traced over time, and it is not the only one that can be predicted by a computer model. Other aspects of climate are subject to change, such as upper-atmosphere temperature, or the temperature in each hemisphere, the sea ice cover in the polar oceans, or the extent of land glaciers, or low-latitude precipitation. If the predictions of the model with regard to one phenomenon fall within the range of measurements, that would be encouraging but inconclusive. If they agree on two counts, scientific confidence would build. If they agree on, say, five counts, then you could begin to use words like "certain."

The process of comparing computer-model results to measurements is called "model validation." The models say that growing levels of greenhouse gases should already have warmed the Earth. Do the measurements confirm the prediction? The fly in the ointment is always the uncertainty range. Depending on which crude guesses are used, the models might or might not agree with the measurements. As a result, there is no unique "yes" or "no" response to the validation question.

In fact, when the guesses in the model are very crude, scientists occasionally adjust them so that the predictions of the model agree with the measurements—in other words, so that it gives the right answer. Physicists call this process "fitting" the data, and it's done in order to find out which of several possible crude guesses is correct. If the latitudes are wide enough in the crude guesses, the computer model can be jiggered so as to agree with any set of measurements. The uncertainties are big enough that one can "fit an elephant." (This is said to be an optimum situation for theoreticians. Economists' methods are similar, and they have a saying, "Models are like sausages: you don't want to know what goes into them.") Yet scientific fudging in this way gets harder as the elephant gets bigger. In other words, as the number of measured properties increases, there are just too many requirements to satisfy and not enough latitude in the guesswork to do so.

Think of trying to cover a mattress with a fitted sheet. Now think of the corners of the sheet as the predictions of the computer, which must match the measured properties, the corners of the mattress. If the sheet has loose corners, then it is easy to fit it to the mattress; there's lots of freedom in how you tuck in the sheet. But as the corners get tighter, maneuvering room shrinks and finally disappears. Eventually there is only one right answer: the sheet either fits the mattress or it doesn't. The model and its predictions are either right or wrong. Likewise, six corners are harder to match than four. As of now, the corners of the models are still relatively slack. In addition, the temperature in the lower atmosphere and the temperature in the upper atmosphere are the only two "corners" of the mattress for which a useful history of measurement exists.

In time, judgment will become easier even if more properties aren't measured or predicted narrowly. The models predict that the Earth will warm steadily. If the Earth fails to warm fast enough, the highest available reckoning of the measured temperature change will fall below the lowest prediction from the model obtained from playing with the internal crude guesses. In this case, the model will have failed the validation test and will be exposed as fundamentally flawed. It could fail for one of two reasons: either our understanding of the effect of added greenhouse gases could be mistaken or other factors, which are not precisely understood and not perfectly reflected in the models—such as solar variations—could swamp the greenhouse effect with an unexpected cooling. The latter is unlikely to occur because we have some idea of the magnitude of the historical changes due to these other factors, such as the slight cooling which particularly affected the Northern Hemisphere between 1940 and 1970. They have probably not exceeded more than about two degrees in either direction (compared with the 1880s temperature) for thousands of years. The measured warming should creep above that range of uncertainty over the next decade or so, if the models are correct. Like suddenly driving into the range of a weak radio station, the "signal" of warming will exceed the "static" of natural variations. Even so, nature has been known to throw curve balls occasionally, so the judgment still won't be definitive.

From this discussion, it should be obvious why a midlevel bureaucrat in the White House's Office of Management and Budget tried so hard to suppress Hansen's May 8, 1989 congressional testimony on a related question. To judge is a normal scientific practice, yet the scientist who does so publicly may have a profound influence on policy.

Compared to other features of climate, the 100-year temperature change is a relatively easy property to simulate with a model, and we have a rich historical data base to match with the computers' findings. There is also far more agreement about the future course of global mean temperature than about other characteristics of the climate. The reason is that calculations of global properties over long time periods, like the world's mean temperature for the 2030s, are subject to far less uncertainty than are estimates of the individual episodes, such as the July 2031 temperature in Nebraska, that make them up. It's like dropping a bag of sugar on the floor: one can be pretty sure how much terrain five pounds will cover, but who knows where each grain will potentially end up! That's why long-term global climate fifty years from now can be predicted with a modest degree of confidence, but next week's weather cannot.

Climatologists are still trying to get a handle on the smaller scale, using their computer simulations. These models are built from equations, like Newton's laws, which approximate the physical behavior of the air itself, when it is heated by the sun, dragged down by gravity, lifted by mountains, and cooled by its own radiation of heat. The result of all of this pushing and prodding is a swirl of air currents, some wound up tightly into storms near the surface, others, like the jet stream, girdling the planet at higher altitudes. Ultimately, this commotion determines the weather.

But what appears as a small kink in the jet stream on a global scale can cause a giant storm to rage at the surface; such kinks are hard to forecast, and, even if one averages over the entire United States, patterns for the future are difficult to predict. It's like trying to match the sheet to each bump and valley in the mattress. As with the global mean temperature, we know that global rainfall will increase when the world warms, because more water will evaporate from both land and sea, and what goes up must come down. But there will be more water in some places, more drought in others because the average is made up of individual storms whose trajectories and intensities will shift in all directions, like the grains of sugar scattered on the floor.

As a result, there remains controversy on such questions as whether the U.S. grain belt will or will not be drier during the growing season forty years from now. At the rate the models are improving, we may not know until we get there. Whatever we find in 2030, the answer could change again shortly as the climate warms more. In short, we face uncertainty, indefinitely.

To avoid giving politicians a confusing array of divergent opinion, scientists often put together committees to rationalize judgments. These attempts to operate by consensus make the whole process social and cultural, and different cultures will render different judgments. A case in point is the 1983 National Academy of Sciences committee, which was swayed by several participants, including economists Nordhaus and Schelling, and agricultural scientist Paul Waggoner, who were all relatively sanguine about the human potential to adapt. Consequently, the report conveyed a think-but-don't-act message, quite different from the EPA study published at the same time. Then, just five years later, with no radical change in scientific understanding, the Academy published a white paper for President-elect Bush prepared by its National Research Council with the aid of outside advisors, including Nordhaus but none of the others who were involved in the 1983 report. This time the Academy tilted toward immediate action.

At the 1987 follow-up to the 1985 Villach conference, William Clark of Harvard's Kennedy School asked a small working group of scientists to assess the validity of the predictions of various climate models. In what range would global temperature lie in 100 years, and with what likelihood? Scientists generally are reluctant to make public pronouncements on uncertain matters and loathe to acknowledge the judgment factor when they do so. The remarkable thing about this evaluation was that the participants didn't hide behind a mirage of supposed facts. In a rare moment of public candor, the scientists declared exactly what they were up to. They gave fifty-fifty odds that the temperature in 2020 would be more than two degrees warmer than today. In so doing, they simply described their findings as being ''the professional judgment of the Villach 1987 experts group.'' The scientists understood that in this instance, their opinion was far more valuable than their certainty.

THE ECOLOGISTS

The Clark group was interesting in another respect. It included three people of a sort who generally shy away from making global generalizations. The three were ecologists. An ecologist is a biologist who thinks about the mutual relations among organisms and their environment, about living systems, such as a chunk of forest or grassland or estuary. An ecologist examines what happens when such a system is pushed this way or that by pest invasions, or by acid rain, or by climatic change. An ecologist might consider how the system pushes back, too, as George Woodwell has in his theory of soil-methane feedback discussed in chapter 2.

Atmospheric scientists have dominated discussions of the global-warming problem for so long that it is often forgotten what the real stakes are: the survival of the biosphere. But as the day draws nearer that politicians *must* make decisions and begin to spend public money on solutions, let us hope that the voices of more ecologists begin to be heard. If any scientists are equipped to assess the fate of humans and nature in the greenhouse world, they are the ecologists.

The trouble with most ecologists is that they learn to love a square meter of soil. In that square meter thrives a microcosm of thrills and chills, of fungi and bacteria, of beetles eating other bugs, of one animal's excreta serving for another's lunch. By and large, ecologists see the world through a kaleidoscope of complexity and think of generalizations as a trap. With this perspective, they are an unlikely bunch to discuss the biological impact of warming on a scale that is politically compelling.

There are certainly exceptions to this rule. There are fifteen Long Term Ecological Research Sites funded by the National Science Foundation in the United States, where the objective is to study the ecology of a large slice of nature, such as an entire forest or grassland, for an indefinite period. For instance, scientists at the Hubbard Brook Experimental Forest in New Hampshire have been studying rainfall, forest productivity, stream-water chemistry, the fate of leaf litter, and other such topics since 1963. Of particular importance are the nitrate levels in stream water.

Nitrate acts as a nutrient, or fertilizer. The forest loves to eat nitrate, which it obtains through the bacterial breakdown of air-borne nitrogen molecules. One class of soil bacteria separates the two nitrogen atoms that form the nitrogen molecule and produces ammonium from them; another class transforms the ammonium into nitrate, a combination of nitrogen and oxygen. This process also releases some nitrous oxide, as we mentioned in chapter 2. The transformations of molecular nitrogen to ammonium and of ammonium to nitrate are called nitrogen fixation and nitrification, respectively. They require considerable energy on the part of the bacteria. If nitrogen and oxygen combined easily, the atmosphere would spontaneously react and there would be no oxygen left to breathe.

When fossil fuels are burned, enough energy is present to scissor nitrogen molecules in the air and recombine the atoms with oxygen. This is one of the processes that creates nitrogen oxides, NO_x (a combination of NO and NO_2) in flue gas or tailpipe exhaust, the other being oxidation of nitrogen atoms in the fuel itself. Nitrogen oxides are rapidly converted to nitrate in the presence of moisture, so fossil-fuel burning always leads to nitrate formation.

Precipitation brings the nitrate to earth in acid rain, where most of it is eaten by growing forests, while the rest simply soaks into the soils and dribbles away in creeks or streams. If a forest isn't utilizing much nitrate, it may be in trouble, like a child with a bellyache who won't take another proffered piece of candy. Then a lot of it shows up in streams and rivers.

Some nitrate is eaten by bacteria in the rivers, which denitrify it, reversing the original transformations and turning nitrate back into molecular nitrogen. Nitrous oxide is a minor by-product of denitrification as well as of nitrification. Any nitrate molecules that avoid the denitrifying bacteria are finally discharged along the coast where they fertilize the algal blooms that periodically choke estuaries. Thus, nitrate in streams is very interesting to ecologists. It is an indicator of problems and can cause more of the same.

But if you ask an ecologist for a theory that explains why trees at each of the fifteen long-term monitoring sites have been eating fifteen different amounts of the nitrate rained upon them, you may get only a blank stare. It's not that a few reasons could not be advanced, it's that ecologists have learned from hard experience that general theories strung together from specific cases are as delicate as a spider web that disintegrates when touched. Instead of risking making an error, most ecologists would rather have no general theory at all.

Scientists do not think much about the philosophy of knowledge these days, but the ecologists' wariness of generalization is directly descended from the empirical view advanced during the Enlightenment by Locke and Berkeley, ending ultimately in Hume's critique of induction. For a modern ecologist, it all comes to this: in drawing conclusions, don't wander too far from the data, or the "ground truth."

Given the complexity of living things and their interactions, and the failure of simple organizing principles, the close adherence to empiricism by ecologists would generally be considered admirable. Yet it may also be argued that now, with the world changing so fast, a little more intellectual risk taking is required of them. After all, an inability to speak in generalizations is a handicap when talking to a politician. Politicians don't want to hear what you don't know; they want to hear what you are confident about, particularly if you are a scientist. The atmospheric scientists who deal with climate are usually physicists of one sort or another, and politicians like physicists because they are usually very sure. Physicists love to generalize. This is their edge, their secret weapon.

THE PHYSICISTS

The story of physics in the twentieth century compares well with the Greek tragedies. It is both enthralling and frightening. The story reveals how much society depends on the judgment of scientists while undercutting the basis of their objectivity with dreams of sugarplums, such as research funds and political power. And it contains critical lessons for the struggle to limit global warming.

If ecologists descend from the empiricists, then physicists descend from another tradition, which is rooted in Plato and Kant. Its essence is self-confidence, of the kind revealed in Einstein's statement, "Nature is the realization of the simplest conceivable mathematical ideas.'' In other words, physicists with their equations and computers can predict things that no one has ever seen.

Einstein was speaking from personal experience. He based his special theory of relativity, which describes the motion of objects at high velocity, on the assumption that the speed of light emitted by a lantern on a railroad platform would be the same whether measured by an observer standing on the platform or by one standing on a train whizzing past. In this way, light differs from a bullet shot by a marksman on the platform, which would appear slower to an observer on a train moving in the same direction. Einstein was brought to this hypothesis because it made the mathematical formulas for the motion of light, Maxwell's equations, particularly simple to solve; and in mathematical simplicity lay reality, as far as he was concerned. His theory had no direct basis in measurement (although some argue that the Michelson–Morley experiment of 1887 must have influenced his thinking).

Einstein published his theory in 1905 and generalized it to include motion in gravitational fields in 1915 (hence the name General Theory of Relativity). One of his predictions, that the path of starlight passing near the sun would be bent by the sun's gravitational field, was not confirmed until observation could be made during the 1919 eclipse. When he heard by telegram of the confirmation of his prediction, Einstein reportedly commented, ''But I knew that the theory is correct.'' When asked what he would have thought if the observation had failed to confirm his theory, Einstein replied, ''Then I would have been sorry for the dear Lord—the theory *is* correct.''

Like Einstein, the climate modelers are physicists who are confident of their theories.* Although there is only a handful of them, and their word is almost all we have to show that climate will warm, the greenhouse effect has commanded attention at the highest level of government. While it remains to be seen whether this attention will lead to political action, it is nevertheless certain that decisive political action will never occur if scientists don't continue to lead.

---

*Confidence in the ''big picture'' presented by the greenhouse theory is tied to experience, including studies of the atmospheres of Venus and Mars and Earth's ice core data. While the models do not adequately predict details, they do reproduce many particulars, like the large seasonal variations in Earth's climate.

As for other public issues, the political will to act on environmental problems generally evolves quite slowly. In the previous chapter, we explored the evolution of the ozone-depletion issue. Take acid rain as another example. The notion that airborne pollutants determine in large measure the acidity of rain originated in England in 1852 with Robert Angus Smith, who blamed the sulfuric acid in the air of industrialized Manchester for corroding metals and fading the colors of dyed goods. By early in this century, trout populations had become extinct in some lakes in southern Norway and had decreased in others, but the link between the decline of fish populations and pollution from distant sources was only forged in 1968 by Svante Odén in Sweden and then ecologist Gene Likens and other scientists in the United States and Canada in the early 1970s. Finally, acid rain became a political issue in North America during the Carter Administration, when the United States and Canada signed a Memorandum of Intent in 1980 to develop domestic air-pollution-control policies and strategies.

Over the next several years, scientists and politicians entwined like vipers. The Reagan Administration persistently stalled the Canadians with excuses about scientific uncertainty, though most reputable experts agreed that acid rain was a menace. All too frequently, however, scientists tend to engender more confusion than clarity with their public pronouncements. Rarely do they give a clear prescription for action because even some physicists regard the interface between science and policy as a perilously complex mine field. Where scientists think of themselves as cautious, the public hears only waffling, which makes political stalling easier.

Simple English can make a big difference. For instance, in 1983 a committee of the National Academy of Sciences announced its findings on the potential effectiveness of reducing sulfur-dioxide emissions in order to ameliorate acid rain. In technical jargon, the key issue at hand was if the atmosphere behaves in "linear" fashion with respect to sulfur dioxide—that is, if cuts in sulfur-dioxide emissions would reduce the acidity of rain proportionately. "Nonlinear" behavior would mean that a 50-percent emission reduction might cut acid rain by less than half, or perhaps by no measurable amount at all, in which case spending money for sulfur-dioxide controls would be pointless.

The finding of the committee was that the relationship was probably linear, but instead of saying so directly, the report said that "there is no strong non-linearity." This phrasing was the classic waffle at its best (or worst). At the press conference at which the report was released, reporters were befuddled, and the door was open for coal and utility interests to flimflam the public again. But when one reporter asked committee chairman Jack Calvert of the National Center for Atmospheric Research what the committee's findings meant, Calvert declared that it meant it was time for industry to "get off the dime" and start cutting sulfur-dioxide emissions. By saying so in simple English, Calvert ended any question about the effectiveness of sulfur-dioxide controls and placed the acid-rain issue squarely in the policy arena.

Despite this and persistent pleas from Canada to act, the Reagan Administration remained adamant. Moreover, members of Congress divided not on party lines but on regional lines on acid rain, with those representing polluting constituencies—the auto industry, coal companies, and big utilities in the Midwest and Southeast—successfully opposing those from the Northeast.

Lobbyists for coal and utility interests, who had the eager ear of the Reagan Administration, spread disinformation to Congress, the press, and the public. Between 1983 and 1987, a utility- and coal-industry front group named "Citizens for Sensible Control of Acid Rain" spent more than $4 million fighting legislation and topped everyone on any spending for any issue during 1986.

New scientific findings on the damaging effects of acid rain continued to come to light. Alarming reports of forest dieback, coming first from Europe and then, by 1984, from the United States, fingered air pollution as a likely culprit. Acid rain was also implicated in the unsuccessful reproduction of Atlantic salmon, striped bass and other anadromous fish species in sensitive spawning rivers ranging from Nova Scotia to Maryland. In the spring of 1988, the Environmental Defense Fund published a report showing that nitrates in acid rain were contributing to the algal blooms, eutrophication, and decline of the Chesapeake Bay and probably other coastal waters as well. Finally, as of this writing, seven years since the Academy report, the administration has proposed an acid rain control plan, and Congress seems to be on the brink of action.*

Why didn't the new findings change the political equation sooner? A threshold had been crossed long before. The public had become inured; any new scientific reports were just so much more bad news for an already sick planet. The decision was already in on acid rain: it's bad. But the politics of solution is difficult, so action on the issue, if inevitable, has been slow in coming.

The pattern for the ozone-depletion issue is similar, but there theory preceded observations. After Rowland and Molina identified the ozone-depleting potential of CFCs in 1974, intensive research followed that generally confirmed the theory, but ozone depletion itself remained too small to observe. Within four years, aerosol use had been banned in a few countries, but another seven years were to pass before the Vienna Convention was signed, as the EPA and other agencies dragged their heels on further curbs while scientific work continued.

The actual observation of ozone depletion changed the course of events and assured a strong Montreal Protocol two years later. Perhaps half a dozen scientists, including Rowland, Harvard's Michael McElroy, and NASA's Robert Watson brought the significance of the scientific findings to the public's attention. From here on, it will be political and economic considerations that determine the schedule for the inevitable elimination of CFCs; new scientific results won't make much difference.

Similar case histories of other pollutants come to mind: DDT, phosphates, PCBs. Each went through three stages. First, due to the efforts of a few scientists, an environmental question arose as an issue of public concern. Then public attention focused on the issue, but limited or no action was taken while both the science and the policy alternatives were assessed. Finally, forceful scientific judg-

*The EDF report, which MO co-authored, and which was prompted by RHB, received the rare imprimatur of page one, column one in the *New York Times* issue of April 25, 1988. Still, it appeared to leave the political landscape more or less unchanged. But a presidential election campaign took place during the hot summer of 1988, and in the face of a public agitated over a string of disturbing findings on the environment, both parties committed themselves to controlling acid rain.

ment, based perhaps on additional evidence, moved the question into the political arena for final resolution.

As an issue, global warming is trapped in the second stage, which began with the various reports and meetings of the mid-1980s, as described in chapter 3. Scientists agree that the world will warm and climate will change as long as the greenhouse gases are emitted in large quantity. They agree that disastrous consequences are possible for society and for the natural world. But they are not yet certain that the greenhouse effect has caused the 100-year warming. Neither are they certain how fast the Earth will warm in the future, nor when dire consequences will become inevitable.

These questions are important. The answers to them could determine, in part, when to act and how much must be done to limit the warming. But scientists will not resolve these questions for a long, long time because their computer predictions will be sharpened very slowly. We are doomed to an indefinite process of assessment and reevaluation of data; and while this process does not preclude political action in the near future, it does mean that the judgment of scientists will exercise a crucial influence on the course of policy.

Not all atmospheric scientists will voice their opinions publicly, but many of them echo Einstein's type of moxie. Near the end of 1988, *Science* published an article asserting that the midwestern drought that summer had less to do with global warming than with other factors. Two weeks later a paper in the journal *Geophysical Research Letters* argued that there had been no warming in the United States over the past hundred years, in contrast to the world as a whole. The *New York Times* played up both stories, as if to provide "balance" to its June 1988 page 1 story reporting the Hansen testimony to Congress. Politicians started looking over their shoulders, but physicist and climate modeler Stephen Schneider scoffed at both sides of the argument. "Do I believe that the [100-year] trend is due to the greenhouse effect? Sure. But can I prove it? No." Does it really matter? Certainly not, he insisted, because future warming will definitely occur. "The future is not based on statistics, it's based on physics." Schneider is Einstein's kind of guy.

This conviction among physicists of the power of pure reason has carried them a long way scientifically. George Woodwell has said that ecologists frequently cannot "get out of their skins." Physicists have at least been able to stretch the bounds of theirs. They have invaded and conquered large parts of astronomy, biology, chemistry, climatology, geology, oceanography, and, yes, even ecology* using that secret weapon, their faith in the power of generalizations based on underlying principles. They can stare down a complex situation and not blink.

Consider Michael McElroy's group in the Department of Earth and Planetary Sciences at Harvard. McElroy was trained in the Department of Applied Mathematics at Queen's University in Belfast, amid a core of brilliant atomic physicists who developed much of the science of the upper atmosphere. In 1971, before Rowland and Molina identified the dangers of CFCs, Berkeley's Harold S. Johnston became concerned that the release of nitrogen oxides in the stratosphere by the proposed SST, the supersonic transport plane, could threaten the ozone layer. Some nitrogen oxides already existed in the stratosphere, the product of a chemical reaction involving nitrous oxide. McElroy began to wonder where on Earth the nitrous oxide came from. (No known process in the atmosphere gives rise to nitrous oxide, so it must come from the Earth's surface.)

As noted earlier, bacteria in rivers turn nitrate into nitrous oxide, regardless of whether the nitrate originally came from natural nitrification, from acid rain, from runoff of fertilizer from farms, or from sewage. So if you want to measure natural sources of nitrous oxide and how humans might add to them, you can't just sit in a laboratory in Cambridge, Massachusetts. McElroy's colleague Steven Wofsy, also an atmospheric scientist with a physical chemistry background, wound up running boats up and down the Potomac and to the mouth of the Amazon looking for nitrous oxide. The results of this work have found their way into so-called "global nitrogen budgets," which generalize from a few measurements to the whole world. It's a necessary chore, and a dirty one, scientifically speaking, and an ecologist wouldn't be caught dead doing it because he knows

---

*Robert May of Oxford, a leading ecologist and one of the field's few theoreticians, started out in astrophysics. May's research follows from the work of A. J. Lotka, who is mentioned in chapter 3, another physical scientist who focused on biological problems.

that every river, estuary, and mud puddle is different. But it could take several centuries to make the ideal set of measurements that most ecologists would feel safe generalizing from. If you are worried over the fate of the ozone layer in the next twenty-five years, you can't afford to wait. That's one reason you find physicists mucking around in swamps.

This incursion from the atmosphere into the biosphere by physicists has been met with considerable consternation by ecologists, who know all about invader organisms. Of course, more is at stake in this case than territory. The issue is cold, hard cash. And here the players are not even close to being equal. After all, no scientists (outside the medical field) know how to squeeze money out of the government like physicists.

THE TRIPOD

Science, politics, and money have supported one another like the legs of a tripod since the early part of this century. The original expectation was of industrial invention and its commercial benefits to fledgling industries like instrument manufacturers; and the National Bureau of Standards (now called the National Institute of Standards and Technology) funded scientific research accordingly. With the advent of modern technical warfare, including submarine and aviation combat in World War I, scientists actively urged an advisory role for themselves. George Ellery Hale of the University of California engineered the creation of the National Research Council within the National Academy of Sciences to dispense technical advice to the executive branch. But funding for basic research failed to materialize after the war, as the government decided to support only those efforts that were directly under military or other government supervision, such as the work done at the Naval Research Laboratory.

World War II again saw the physics community spearhead the organization of a scientific effort; this time, the physicists made sure that they were holding the brass ring tightly when the last shot was fired. When Vannevar Bush of M.I.T. and the Carnegie Institute put together the National Defense Research Council, he funneled most of the funding to corporate and university facilities, such as M.I.T.'s Radiation Laboratory, which developed radar systems. The Manhattan Project, which culminated at Los Alamos, relied on fission and isotope separation studies at Columbia, the University of Chicago, and Berkeley.

Taking advantage of the perception that research scientists could deliver the goods, military and otherwise, Vannevar Bush and others cemented an ongoing university-government partnership. Funding for basic physics research was split between the independent National Science Foundation and the military. As Daniel Kevles wrote in his book, *The Physicists,* "The Los Alamos generation [of physicists] insisted that if it was the responsibility of the civilian scientist to contribute his expertise to defense research, it was also the responsibility of the federal government to finance the basic research and training on which the national security ultimately depended."

Military research is not the sole source of support for physics, and ecology is not bereft of government funding. The point is that physicists learned how to turn a societal need into an annuity for the discipline. In the process they provided support for an entire economic and political agenda, which included not only the Bomb, the military budget, and the Cold War, but also lasers, nuclear power, and outer space.

The Manhattan Project was much more than an early example of the government-science complex: it became something of a creation myth for modern physics. As a group, physicists resemble religious disciples a generation after the miracles. Their titular god is Einstein, and their monthly magazine, *Physics Today,* still features faded photographs of him and other greats as it bemoans a long-gone sense of purity. The myth derives from several sources: the heroic nature of the task, a sense of camaraderie never to be reattained, the sudden elevation of a small community of innocents to political significance, and finally the loss of innocence in what some regarded as a tragic outcome. But don't be fooled by their lingering unease. Physicists are protean. Just as the miracles of particle physics and space travel are losing luster, they are beginning to metamorphose into saviors of the global environment.

Watching physicists dance before Congress in this new role can be like an evening at the Bolshoi. Even atmospheric scientists, far afield of the nuclear physicists, seem to twirl into the political arena as if by birthright. One of us testified at a Senate hearing on the ozone hole in the fall of 1987 that featured several classic performances. The seven witnesses, six of whom were atmospheric scientists, included McElroy and James Anderson of Harvard, who had designed the definitive experiment that pinned the blame for the ozone hole on CFCs. His measurements endowed the issue of ozone depletion with a clarity that is common in laboratory physics, rare in environmental experiments, and unknown in ecology. His work left no doubt that CFCs caused ozone depletion in general, and the ozone hole in particular. And the senators had what they wanted: scientific certainty.

Anderson's masterful presentation was followed by a typical McElroy appearance: in 1984, he had published a paper on the role played by marine phytoplankton in the removal of carbon dioxide from the atmosphere by photosynthesis. Now he speculated that the ozone hole could permit so much ultraviolet radiation to penetrate to the polar ocean that the phytoplankton would be destroyed. Inasmuch as coldwater phytoplankton can suck carbon dioxide from the atmosphere and deposit the carbon on the ocean floor, McElroy reasoned that damage to the ozone layer might accelerate the greenhouse effect. McElroy had done the senators favor number two: he had given them a new global connection.

The session concluded with McElroy, Anderson, and the others reminding the senators that without fifteen years of grudging research support by government none of them would have been able to perform and interpret these elegant experiments. How would they, the decision makers, then have known what to do? The scientists proceeded to describe future experiments to the receptive senators, which might cost millions of dollars but which might help save humanity from its own stupidity.

Scripting the same scene with six ecologists instead of atmospheric physicists would be akin to casting Woody Allen instead of Gary Cooper in *High Noon*. The ecologists would probably not have a definitive experiment to describe, or a unique explanation to advance. Neither would they be able to propose a set of experiments that could elucidate global questions in less than a decade.

Nor would they be likely to advance a theory of how a problem in Antarctica could result in a cataclysm in some senator's home state.

Humility has its virtues, however, and the physicists' brand of clarity and their sense of mission have less attractive aspects. The proposal for the Superconducting Super Collider (SSC) project, a $6 billion ring-shaped particle accelerator that appears to be the last gasp for big-time nuclear physics in the United States, has been pressed vigorously on the Congress and the Administration by several physicists, most notably Nobel Prize winner Leon Lederman. Its future is highly uncertain because the physics community itself is riven with doubt about the value of multibillion-dollar projects while small-scale research goes begging.

The notion that the SSC project would yield $6 billion in scientific dividends is nonsense. Indeed, it is doubtful that its congressional sponsors are fooled by such claims. Rather, the future of the project rests on two other formidable arguments: concerns about national security, and the possibilities for political pork barrel. The site chosen for the project by the Department of Energy is Texas, the home of George Bush, Lloyd Bentsen, James Baker, former speaker Jim Wright, and other political luminaries. More importantly, the SSC is a vehicle for U.S. leadership in high-energy physics, and high-energy physics is a potential source of scientific talent and technology for weapons-development programs. Even if the SSC is never approved, the fact that it has gotten this far in a budget-slicing era is testament to the potent combination of science, money, and politics, particularly national-security politics.

But that potency is now flagging. Military expenditure in general is under assault as the Soviet Union's recent actions are making the Cold War harder to fight and other needs press for attention. The big science–military connection received a black eye from the deceptions that accompanied Star Wars (much to their credit, several hundred physicists publicly refused to participate in related research). Perhaps most significant, the older generation of Manhattan Project leaders has very nearly expired. Edward Teller is aging; Richard Feynman, I. I. Rabi, and many others, on both sides of the argument over science and the military, are gone.

Furthermore, there will be no new generation of nuclear leaders endowed with the mythic aura of the Manhattan Project. Although military-oriented research and development still drains away vast quantities of money and talent, the weapons program itself no longer attracts top talent. As one Department of Energy official commented during the Fernald–Hanford–Savannah River weapons reactor mess in 1988, when the public discovered that the Department of Energy couldn't keep its own bomb factories operating, "It is next to impossible now for us to attract good quality people and very hard even to keep the good young people we have."

THE NEW ARISTOCRACY

A new concept of security based on environmental integrity is gathering strength. Atmospheric physicists have led us to grasp the seriousness of global warming, but they cannot lead us out of the problem on their own. The charge must be joined by ecologists. For something larger than intensified hurricanes, drought, and heat is required to focus continued attention on a problem that, like ozone depletion in 1978, has not fully arrived. What is needed is a knockout punch. Because warming and its consequences lag emissions by decades, they must be understood to threaten the continuation of life on Earth or no one will pay attention until things get out of hand.

Who can deliver that punch? Who can write a letter equivalent to Einstein's to President Franklin D. Roosevelt about the threat of Germany's developing the Bomb? Not the atmospheric scientists. For all their venturesome spirit, they cannot make judgments about life itself. It lies too far outside their discipline. But the ecologists can. It remains unclear, however, whether they will be willing to do so.

Following a 1988 scientific conference in London on ozone depletion, James Lovelock of Gaia renown, NASA's Bob Watson, and ecologist Robert Worrest, chief of EPA's research on ultraviolet-radiation effects, met with the press. A reporter asked them about the significance of large-scale ozone depletion. Worrest used the word "disaster" to describe the consequences of a 50-percent global depletion, then thought better of such an extreme remark and began to temporize. No one in the room would respond directly to the question of when ozone depletion might turn from nuisance into disaster.

It's a question for ecologists and it's a matter of judgment: after all, life can exist with the ultraviolet radiation that would result from halving the current ozone levels. For instance, the tropics receive three times as much ultraviolet radiation as do the latitudes of the northern United States, yet at least some northern species, such as fair-skinned people, can survive in the tropics (albeit with higher skin-cancer risk). The question is what the Earth as a whole would look like, how the relationships among species would change, whether weeds would choke out trees or viruses run rampant in humans.

But consider what can happen when ecologists and climatologists sit down together to predict our biological future. An ecologist might say, "Predict the climate for my square meter in 2032, so I can predict what will live there," to which a climatologist might retort that he can barely produce a precipitation scenario on a continental scale. How can the distance between the two disciplines be closed? Climatologists must be willing to do what William Clark's working group at Villach did by discussing probabilities and ranges of future climate. The ecologists must be willing to emerge from their square meter and hazard a few generalizations.*

Ecological ideas have moved like a wave through the United States in the past, starting with George Perkins Marsh, followed by Theodore Roosevelt and Gifford Pinchot's conservation movement, the land-conservation effort spawned by the Dust Bowl, and most recently the environmentalism of the early 1970s stimulated by Rachel Carson. Along the way, a few individual ecologists and biologists have provided leadership, including most recently George Woodwell, Gene Likens, Barry Commoner, Paul Ehrlich, and Peter Raven. The pesticide DDT may have become a public issue at first

---

*A small effort is now under way. It is called the International Geosphere Biosphere Program or, in the United States, Global Change. It is led by a biologist, Stanford's Harold A. Mooney. This is a small step in the right direction. Global Change is really a mongrel science searching for a philosophy. Under its umbrella, ecologists may yet develop a global view; but as of 1989, the IGBP is too starved for funds to get started properly.

because of Rachel Carson, but its use in this country was halted by the public efforts of marine ecologist Robert Risebrough, biologist Charles Wurster, Woodwell, and others in the Environmental Defense Fund. "Biological diversity" and "nuclear winter" are familiar phrases in large measure because of the activities of Raven, Woodwell, and Ehrlich, while Likens has made acid rain a household word. But these are isolated examples. Daniel Worster, speaking from the perspective of the 1970s environmental movement, saw the ecologist in a political sense as "mediator between man and nature." Unfortunately, a genuinely organized effort by ecologists, fighting for a research agenda and for the environment on a continuing basis, never coalesced.

As a result, the creation and propagation of environmental issues in the public arena is dangerously dependent on the handful of scientists who, like Hansen, Woodwell, and Gordon MacDonald, have made it their business to push them into the spotlight. An inspection of the chronologies of either the greenhouse or CFC issues in chapter 3 shows how only a few people carried the lion's share of each effort.

In the early fall of 1987 we had lunch with a brilliant and broad-minded ecologist, seemingly the type of fellow to lead his minions out of their square meters. We discussed the possibility of forming a steering group to interpret the consequences of global warming. Each name suggested brought a grimace or a grumble from the ecologist: this one was a charlatan, that one had stolen his ideas, the other one was a pain in the ass. Physicists are relentlessly personal in their opinions, too, but they know when to put them aside. As Robert "Skip" Livingston, an outspoken marine ecologist at Florida State University, says, "Physicists usually work together, but ecologists indulge in internecine battles. They put up their banners, challenge one another to battle, and then retreat to their little castles shouting victory."

It is time for a different approach. It took the physicists two world wars and fifty years to get what they needed and to give the world what it thought it wanted from them. Ecology is a young science, but the world cannot wait fifty years this time. Ecologists should consider the current situation as preparation for war and, along with other scientists, organize the way the physicists did earlier.

This proposal for eco-activism will be met with skepticism by some scientists, who fear that to speak out publicly risks the embarrassment of error, and who warn that the physicists' pursuit of power and funding has sometimes led to the sacrifice of scientific objectivity, and worse. But such instances in history should serve as cautionary tales, not obstacles to involvement. Who else but the ecologists can tell us what makes the Earth's heart beat and how we can keep it going? In the debate over global warming, who else will speak for the Earth?

CHAPTER 5

Over the Cliff:

Why Act Now?

DURING THE LONG, HOT SUMMER OF 1988, TED KOPPEL OPENED AN ABC "Nightline" program on the greenhouse effect by asking one of us, "I'd love to be able to say to you that I think the American public can get energized over some perceived threat forty years down the road, but I don't believe it. Do you?"

Koppel obviously didn't think the American public had absorbed the lesson that Wile E. Coyote, the "Roadrunner" cartoon character, learns the hard way when he races along a road that, unbeknownst to him, is about to end at the edge of a cliff. Wile E. Coyote moves so fast that he's ten feet past the sheer dropoff when he freezes, looks down, and, in shock, comprehends his situation. For a split second, while suspended in midair, he flails his arms in a futile attempt to reach the cliff. Then gravity takes over, and he plummets out of sight. It is worth bearing this image in mind as we explore the reasons for applying the brakes to global warming sooner rather than later.

The 1980s seem to have thrown up one obstacle after another to focusing on distant threats, and many of these problems remain unsolved. The economies of Brazil, Mexico, Venezuela, and Nigeria all stumbled as oil revenues shrank and their debt piled up. The Iran–Iraq War and the Soviet invasion of Afghanistan lasted throughout most of the decade. The contras fought the Sandinistas in Nicaragua, death squads roamed El Salvador, Lebanon was torn to pieces, Palestinians in the West Bank erupted against Israel, and South Africa threatened to explode. Internecine warfare continued in Northern Ireland and India, while terrorists seized hostages, boarded cruise ships, and blew up jetliners. The AIDS epidemic ran unchecked in central Africa and parts of the West, China erupted briefly with an insurrection of democracy, and the political landscape of eastern Europe changed overnight. With so many immediate life-and-death decisions to be made, why should political leaders take time to ponder the seemingly remote hazards posed by the greenhouse effect?

An optimist might note that Margaret Thatcher, Mikhail Gorbachev, and George Bush each have spoken out on global warming. Already, the United Nations has passed resolutions, and governments have set up committees. After thirty years of intermittent concern, the public in the United States and Europe is once again agitated about the environment.

But when it comes to making technically straightforward but politically difficult decisions, such as restricting automobile use or coal burning, politicians, at least in a democracy, usually waffle, wiggle, and head for the hills. Most of them cannot see beyond their term of office, be it two years, four years, six years, or even longer. In general, outside Scandinavia, credible governmental leadership on environmental issues is notoriously deficient. Until devastating consequences strike, global warming could become just another unsolved problem on a very long list of environmental headaches and, with the real heat decades off, not a very urgent one at that.

In a logical sense, the arguments for immediate action are compelling, and they arise from both the physical nature of the problem and the customarily protracted pace of government, even once it has decided to move. A case in point is the glacial response to acid rain in Europe and North America.

Although the United States is on the verge of enacting an acid rain control program (see chapter 4), a full decade will have elapsed since the Carter Administration first acknowledged the need to limit emissions, and fifteen years will have elapsed since acid lakes were found in the Adirondacks. Ten more years will elapse before any acid rain legislation is fully implemented.

The evolution of the issue in Europe was similar. Complaints by Sweden and Norway since the early 1970s, that pollution from Great Britain, West Germany, and the East Bloc was destroying tens of thousands of lakes, fell on deaf ears. Then, in the early 1980s, outrage over dying forests in West Germany swelled the Green Party's ranks until they eventually garnered 9 percent of the national vote. Forest decline created a major issue in the 1983 West German elections, and by March 1984, Germany and nine other countries had agreed to cut sulfur-dioxide emissions by 30 percent. By 1988, the same countries had decided to freeze nitrogen-oxide emissions and the European community was undertaking automobile-emissions controls for the first time. West Germany and two other countries were moving toward a 60-percent sulfur-dioxide reduction scheduled for 1998.

With this sluggish process in mind, let us consider the even greater need for immediate action on the greenhouse effect. It is the result of two menacing features, demons if you will, of global warming: irreversibility and the lag time between emissions and effects. These characteristics distinguish global warming from other environmental issues, and they have the vicious consequence of increasing the need for an urgent response while at the same time making it politically difficult to implement one.

IRREVERSIBILITY

As long as greenhouse gases are emitted in quantities close to current amounts, Earth will become warmer and warmer for an indefinite period lasting at least hundreds of years. If emissions increase continuously as they have in the past, warming will accelerate. Should emissions be reduced, greenhouse-gas levels still would remain elevated for centuries, making their consequences irreversible in any human time-frame.

To understand why, consider a bathroom sink. With the drain partly shut and the water running fast, the level will rise and the sink will overflow. Conversely, with the drain wide open and the faucet almost shut, the sink won't fill at all. But if both the faucet and the drain are wide open, or both are almost shut, or both are somewhere in between, the sink will fill to a certain depth and that level will hold. This level is called the steady state, and it remains constant because the inflow from the faucet is exactly matched by the outflow from the drain. There is a balance.

Another property goes hand in hand with the steady-state behavior: reversibility. For example, if the faucet is opened a little bit more, the water level will begin to increase but the pressure of the extra water will also force an increase in the flow into the drain. The depth of the water will increase until the higher outflow exactly matches the increased inflow, and a new steady state is reached. If the faucet is then tightened to its original position, the water level will begin to drop; but the pressure will also drop, so the flow out the drain will begin to slow down. The water again reaches the old steady-state level. The change was reversible.

Most air pollutants possess this steady-state property. As long as chemical emissions remain more or less constant, the amount in the atmosphere will vary little once they reach their steady-state level. If we want to reduce their level in the atmosphere, all we need to do is cut emissions. For acidic pollutants like sulfur dioxide, the faucet is smokestacks and the drains are rainstorms, which wash the acids to the ground, and leaf surfaces, which filter them out as they blow by on the winds. In this particular case, the drain is wide open, for no more than a few days of emissions ever builds up in the air. Consequently, pollution levels are reversible. If emissions were cut to zero, rain would scrub the air of sulfur dioxide within a few days.

The case of urban air pollution provides an example of how dependent political thinking has become on the expectation of reversibility. Between the end of World War II and 1970, smog levels increased steadily as automobile use grew across the country and smokestack pollution remained largely uncontrolled (except for soot removal). In 1970 and again in 1977, the Clean Air Act of 1963 was significantly strengthened, and this sharply reversed some of these trends: new power plants began to burn low-sulfur coal or to employ scrubbers, while new cars carried catalytic converters. Both new cars and new power plants emitted some pollution but considerably less than the older ones. As older cars and power plants were retired, the air improved.

But now all existing cars have tailpipe controls of one sort or another. Even though these controls have been tightened over time, each expansion of the country's fleet adds a bit of pollution. In 1970, there were 90 million passenger cars in the United States; in 1987, the total was 140 million, an increase of 50 million cars in only seventeen years, which is now erasing improvements anticipated from tightening controls. Like new cars, each new power plant adds a bit of pollution; and because utilities have stretched the lives of their old plants, these bits cancel the benefit from the few that retire. Thus improvement in air quality has long since ceased, and pollution levels have remained steady for the past decade. The amount of pollution goes up and down with the vagaries of the weather and the economy; but, overall, little progress has been made.

Nevertheless, the cleanup that occurred in the first years of the Clean Air Act amendments means people no longer drop dead suddenly, obviously, and in large numbers during severe smog episodes, as they did in Donora, Pennsylvania, in 1948, when 6,000 out of 14,000 people became sick and 17 died; or in New York City, where perhaps as many as 400 people died in 1963 and another 168 died in 1966. As a result, any further strengthening of the Clean Air Act has been stalemated for a decade. As in the case of acid rain, which could be reduced with controls on the same set of pollutants, the general public did not become sufficiently exercised until the late 1980s to force politicians into action. One basis for complacency has been the belief that the situation is reversible. If, at any time, the decision is made that continual damage is unacceptable, smog and acid-rain levels can be lowered as soon as emissions cuts are enforced, because these pollutants do not remain in the atmosphere very long after their emissions are reduced.

Yet the situation with global warming is entirely different. Let's return to the analogy of a sink with both faucet and drain equally open. Now consider what happens when the drain is tightened until it is only open a crack. If the faucet is left in the same position, the water level will begin to rise. Even if the faucet is turned down somewhat, the level will climb. How fast it rises depends on how

79

much the faucet is opened, but in any event the climb will continue until a new and much higher steady water level is reached. If the drain is shut completely, the water level will rise forever, onto the floor, out the door, no matter how slowly the water runs.

Unfortunately, greenhouse gases resemble water in a sink with a nearly closed drain and a wide-open faucet. Much more of these gases is emitted into the atmosphere by human activities than the drain can accommodate. Thus, the level of greenhouse gases in the atmosphere keeps rising. In the case of carbon dioxide, the drain is the ocean and the forests, which can absorb only half of each year's emissions from the atmosphere. The process of removing the remainder is so gradual that if all human carbon-dioxide sources were eliminated today, the extra gas already accumulated would remain airborne for more than 300 years. Compared to the faucet, the drain is nearly shut tight, so the carbon-dioxide buildup is effectively irreversible.

But the situation is actually even worse, because emissions of carbon dioxide have been increasing. We keep opening the faucet more and more, and carbon dioxide keeps rushing into the atmosphere at ever greater rates of accumulation. In contrast, we would have to close the faucet by more than half, cutting emissions by more than 50 percent, to keep the gas level from climbing and stop the warming.

What is true for carbon dioxide is also true for two of the other greenhouse gases, the CFCs and nitrous oxide. The typical CFC-12 molecule can avoid its fate, destruction by ultraviolet radiation, for more than 100 years; the nitrous oxide molecule can avoid its fate, destruction by chemical reaction, for about 150 years. This long lifetime, or drainage time, means that about six times as many CFCs are emitted in a year as can be drained from the atmosphere.

Continuously higher greenhouse-gas levels must translate into an ever-warmer Earth unless negative climate feedbacks eventually come to dominate the positive ones. But there is no compelling evidence to suggest that they will. As emissions grow, the buildup of greenhouse gases, and Earth's consequent warming, could continue without limit until the sources, like fossil fuels, simply run out. Unfortunately, there are enough recoverable coal reserves on Earth so that warming could run well into the double digits before the coal is all used up, sometime in the twenty-third century.

There is, however, a man-made solution that might be employed to reverse the buildup. If humans can accelerate the part of the carbon cycle that puts carbon dioxide into the atmosphere, then perhaps they can accelerate the part of the cycle that removes it. It could be done by expanding the global rate of photosynthesis. Due to deforestation in the tropics, the Earth's forests, on balance, may be feeding more carbon dioxide into the atmosphere at this time than they are absorbing. But deforestation could be slowed and fast-growing trees, which mature in about a human generation—twenty-five to forty years—could be planted over large areas, continuously storing carbon as they grow. The task would be monumental, requiring an arable land area only slightly smaller than the entire United States just to compensate for the next forty years of emissions.

With global cooperation, this scale of reforestation might be possible. Appropriate land could be made available, albeit not in one place. But the complexities of reaching such a global accord make trees a slender reed to lean upon for an ex-post-facto rescue. Moreover, if we waited too long, the climate might keep changing so quickly as to challenge our ability to find places in which particular kinds of trees could survive and grow even for a few short decades. Viewed as a quick means to reverse warming after the fact, reforestation would probably fail. But reforestation makes sense if it is viewed as a means to slow the buildup of carbon dioxide by a modest amount starting now, while other solutions are implemented.

Variants on the reforestation scheme have been proposed. One approach would be to encourage algal growth in the open sea or on artificial ponds, which would take carbon dioxide from the atmosphere. But it would be dangerous because algae themselves emit other troublesome gases.

The situation is even more problematic for CFCs and nitrous oxide. No one has yet devised a credible method for accelerating their natural rate of drainage from the atmosphere. One futuristic scheme envisions building lasers at the tops of mountains to blast CFCs, in an environmental version of Star Wars. Proposals like this one have always failed because of their cost or energy requirements; moreover, they could lead to unintended and drastic consequences.

Other fantastic schemes have been proposed to counteract the warming effects of the gases, instead of halting their increase in the atmosphere. One involves using Earth-orbiting mirrors to cast sunlight back into space, while another calls for shooting rocket-borne sulfur-dioxide canisters into the upper atmosphere. The resulting sulfate particles would also reflect sunlight. But the sulfate would do its reflective work in the stratosphere, and there it could have an effect similar to that of the stratospheric cloud crystals over Antarctica, accelerating the destruction of the ozone layer. (The sulfate proposal was advanced by Wallace Broecker before the ozone hole was understood.) In addition, even small changes in reflectivity would have large and unpredictable effects on climate, photosynthesis, and the atmosphere's chemical balance. The idea that a large array of orbiting objects or particles could be fine-tuned to our needs is nothing short of a pipe dream.

LAG TIME

The full extent of the warming that will accompany any particular level of greenhouse gases will be realized about forty years after their release into the atmosphere. To understand why this is so, consider two Thanksgiving turkeys, one stuffed, the other not. Put the birds in separate ovens and turn the dials to the same setting. The stuffed turkey will take far longer to cook than the empty bird because the stuffing is slow to warm, and it keeps the rest of the bird relatively cool. Still, if the turkey were left in the oven long enough, its temperature would eventually reach the level of the oven setting and stop rising.

Think of the atmosphere as the turkey, the ocean as the stuffing, and the level of greenhouse gases in the atmosphere as the oven thermostat. Without the ocean, the atmosphere would heat quickly, like the empty bird. Add the ocean and the warming rate slows, because heating the great mass of water in the ocean takes a very long time. (For similar reasons, coastal areas or islands remain cooler than the mainland, particularly early in spring.) If atmospheric levels of greenhouse gases are allowed to increase continuously as they are now, it is as though the cook kept turning the thermostat up higher and higher: the stuffed turkey would always lag behind the oven setting. Conversely, if emissions of greenhouse gases were reduced so much that a steady state of the gases' levels was achieved at today's levels, it still would take the Earth's temperature decades to level off after the gases were stabilized.

Unfortunately, the Earth's thermostat is already set rather high. If by some miracle we could stabilize the amount of greenhouse gases in the atmosphere immediately, the global mean temperature would likely climb at least another degree or two. Surely it will take decades to implement a response fully, during which time the thermostat will move higher and higher.

Lag time and irreversibility cut the legs out from under the politicians' responses of "Let's see how bad things get before we spend any money" or "We need more research." Any time our leaders decide to control emissions, there will always be more warming in the pipeline; and they will never fully be able to anticipate its consequences because the warmer world is largely terra incognita. Computer simulations give a broad indication of future climate but say little about nature's response and even less about people's. Ice-core data might warn us of changes that have occurred in the past but which are not captured by the computers, like feedbacks affecting the ocean's currents. Ice cores reach back 160,000 years to a time before the dawn of modern man. Unfortunately, the world was never more than three degrees warmer during that period than it is now. Paleoclimatic data, such as tree rings and fossilized pollen, provide a fair understanding of the effects of *gradual* climate change on living things, but the picture grows dim beyond the past few tens of thousands of years. The analysis of how living systems will respond to larger and faster global warming is as much art as science. The shroud of uncertainty will not be lifted until after the fact, and by then the situation will be effectively irreversible.

These two characteristics put policy makers in an unaccustomed position. Take one who has been told for years that the Earth is warming but that the risks can never be completely assessed. He or she knows that without action, the greenhouse effect will build indefinitely and irreversibly. Yet even if he presses for emission reductions, the climate will continue to warm due to the two demons, and the public will see no immediate benefit. This is a patently different situation from others he has faced in the past, when the consequences of a decision were obvious within a relatively short time. It took less than a decade for the public to see benefits from the Clean Air Act in the form of lower pollution levels in many cities. But with greenhouse gases, atmospheric response will drag out over many decades, and the situation will get worse before it gets better.

Earth's temperature hasn't quite reached the point that people notice the warming. Without a happenstance string of hot summers, it could be decades before the dire consequences are made absolutely clear. In the meantime, a politician will be inclined to do little or nothing.

But if politicians respond to global warming in this way, as they did to acid rain or ozone depletion, waiting a decade for dramatic discoveries and then taking another fifteen years to eliminate the cause, the planet could be in big trouble. When the theory of ozone depletion was first proposed in 1974, no one doubted the risks associated with allowing the ozone layer to decay. But industry resisted regulation and governments did little. Strong measures might never have been taken without the observations of serious ozone loss, first in Antarctica, then globally. We were lucky. The unpredicted occurrence, the surprise, took place at the bottom of the world. The climatic equivalent of the Antarctic ozone hole, for example, a four-year drought, could happen anywhere, even North America itself.

THE THREE-DEGREE CLIFF

Warming beyond three degrees, the boundary of experience for the modern human species, is like going over a cliff with little notion of how far we will fall. Circumstances are changing so rapidly that, even without knowing it, we may approach the edge in a few short years. The faster we emit greenhouse gases, the further we will be committed to an overshoot before much can be done, and the harder the fall will be when the effects are manifest. As with Wile E. Coyote, we will have gone quite far past the edge before we understand our situation.

Early in the next century, global temperature could be surging at the substantial clip of one-half to one degree per decade, far beyond what civilization has ever experienced on a sustained basis. By then, the world will be about a degree warmer than it is now, and the scientific community will presumably have settled the argument over whether the warming is due to the greenhouse effect; but if nothing has been done in the meantime, the lag factor will guarantee that an additional two-degree warming is already in the pipeline.

If the standard, leisurely posture toward environmental problems is assumed at that point, it could take thirty or forty years to make political decisions, to utilize the existing technologies fully, and to develop and implement new ones, which would allow a big reduction in fossil-fuel use. So one might figure that another three or four degrees of warming could be built in before the greenhouse-gas levels are stabilized.

Thus, if we treat global climate change like any other political problem, the mean temperature of the planet can be expected to reach six or seven degrees above the current level. We will be committed to record rates of warming and unprecedented dislocation. When it was last that warm, humans were still millions of years in the future. A seven-degree rise is probably enough to guarantee the disintegration of the West Antarctic ice sheet and an eventual eighteen-foot rise in sea level. A seven-degree warming is barely within the range of reliability of the climate models, so we have no idea what other shocks might loom in that new world.

By comparison, another degree or two of warming is inevitable as the temperature of the oceans catches up to the present greenhouse-gas level. If we begin to act today, we might achieve everything that needs to be done to hold the global average temperature to a mere three degrees above what it is now. The difference between three degrees and seven degrees is the difference between the edge of the cliff and the bottom of the abyss, from which there is no return.

CHAPTER 6

Heliotrope:

Cooling the Earth with Solar Energy

Once the essential properties of energy are understood, it is readily apparent why we do not really need fossil fuels in the long run and why, in fact, the planet would be better off without them. Even so, fossil fuels will be difficult to abandon because they have become an addiction. As with drugs, whose short-term pleasures cannot be separated from the long-term poisoning of the system, the benefits of fossil fuels inevitably entail those fatal flaws, among them carbon and sulfur. Readily transportable by pipeline, ship, rail, truck, or three-gallon can, easy to store and to transform to other forms of energy, fossil fuels can deliver high densities of power, a surge of energy at one place at a given moment.

The question is, how can we replicate the desirable properties of fossil fuels while avoiding the detriments associated with them? Are there energy sources that can deliver the benefit without the greenhouse poison?

There are only two proposed candidates for this role: solar energy and nuclear energy. Unlike fossil fuels, both are virtually inexhaustible, and neither produces greenhouse gases. But complete replacement of fossil fuels at a reasonable cost by either one would require technological advances. When their future potential is compared in this light, solar power is the only choice.

SOLAR POWER

Sunlight carries energy at low-power density in nontransportable form. At any one place, it is unavailable a great deal of the time and cannot be stored directly. A less flexible form of energy is hard to imagine. Yet these problems can be overcome by converting solar energy to other forms and storing it in substances such as hydrogen. And the major advantage of solar energy is that it creates *no* acid rain, *no* smog, and *no* hazardous radiation. Indeed, solar energy promises a nearly complete cure for all of the faults inherent in fossil fuels.

The conversion and storage of solar energy takes place naturally every day. It occurs when sunlight reaches the surface of the Earth and heats it; it occurs during photosynthesis, when energy is stored in plant life or biomass; and it occurs indirectly when the sun's heat evaporates water from the ocean and lifts it high into the air: the water vapor descends as precipitation, forming rivers from which 7 percent of the world's energy is extracted as hydropower.

Artificial conversion of sunlight to electricity is likely to provide the first step in satisfying energy needs without fossil fuels. Luz International, Ltd. operates the world's largest solar-power plant, in the Mojave Desert. There electricity is produced by concentrating sunlight on tubes that contain oil. The hot oil is circulated through water to produce steam, which is used to run a generator just like the steam produced at a coal-burning plant. While the sun is shining, the generator provides nearly 200 megawatts of power, equivalent to the output of a modest-sized fossil-fuel plant, to customers in southern California, at about twice the cost of coal power but for considerably less than the average nuclear plant.

Although Luz is still expanding, its technology, called solar thermal electricity generation, may not be the wave of the future. The direct conversion of sunlight to electricity by a process called the photovoltaic effect promises to be simpler and, in the long run, cheaper. By eliminating the intermediate step of turning a generator with a steam turbine, direct conversion offers the simplicity and easy maintenance of a system with few moving parts.

The idea that light could be converted directly into electrical energy dates back 150 years. The French scientist Alexandre Edmond Becquerel is credited with first demonstrating the photovoltaic effect in

1839. It is ironic that his son, Antoine, shared the 1903 Nobel Prize in physics with the Curies for the discovery of radioactivity in uranium, the basis of nuclear-fission energy. That was also the year that Svante Arrhenius, who theorized that global temperatures would increase with growing carbon-dioxide emissions, won the Nobel Prize in chemistry for his work on ionization.

Photovoltaic (or solar) cells are slivers of semiconductor, usually silicon, mounted on metal, glass, or plastic. The silicon is "doped" with impurities like boron and phosphorus, which create a permanent electric field within the cell. When sunlight strikes a photovoltaic cell, its electrons absorb the energy and leap free of their parent atoms. Instead of reattaching to the silicon, they are propelled by the electric field, forming a current in the wires attached to the cell.

Before 1954, solar cells of selenium could convert only about one percent of incident sunlight into electricity. But in that year, three Bell Telephone Laboratory scientists—D. M. Chapin, C. S. Fuller, and G. L. Pearson—found that the semiconductor material used in transistors would convert sunlight to an electrical current with at least 6-percent efficiency.

Photovoltaic cells found an immediate client in the space program, which needed a simple, lightweight power source for satellites. Since then solar cells have been used for small-scale devices like pocket calculators. In conjunction with storage batteries, they provide electricity for homes not connected to utility power grids and are particularly useful in the Third World for remote power needs such as water pumps. Solar electricity can also be fed directly into power grids, making more energy available at midday, when factories, offices, and air conditioners are all humming away. During these peak demand periods, utilities must stretch their so-called base load generating capacity by drawing from costly sources, like natural-gas turbines, that they otherwise leave idle. The only factor that inhibits the wider use of solar cells is the cost of manufacturing them.

Under optimum conditions, say in the California desert, a photovoltaic power plant constructed from an array of today's commercial solar cells could produce electricity costing about thirty cents per kilowatt-hour, which is still two and a half times the price of peak fossil-fuel power in that state.* To put it another way, it would cost five dollars a month to light a hundred-watt lamp for six hours a day with photovoltaic energy, as opposed to two dollars with standard power. Although sales of photovoltaic cells have increased greatly since 1975, and prices have come down by a factor of ten, the number manufactured each year in the entire world could only produce one-thirtieth of the electricity generated by a single average U.S. power plant.

At present, a handful of U.S. companies manufacture solar cells, including ARCO Solar,† which has built a 6.5-megawatt photovoltaic power plant at Carissa Plains, California, now providing power to Pacific Gas and Electric, and Chronar Corporation, which is planning one to provide power to Southern California Edison.

Silicon cells are of three types. Amorphous cells are made by spraying silicon vapor to a thickness of a thousandth of a millimeter on a glass, plastic, or stainless-steel substrate. Single-crystal cells are produced by slowly growing a cylindrical crystal of silicon, then slicing it into wafers about a tenth of a millimeter thick. Polycrystalline cells have characteristics intermediate between the other two types.

Two factors determine the high price of photovoltaic power: the efficiency with which a square foot of cells converts sunlight into electricity and the cost of producing that square foot. As might be expected, the cheaper amorphous silicon cells are relatively inefficient, while the more efficient single-crystal cells are more expensive to produce. Progress will come either by improving efficiency, or by reducing production costs per square foot, or both. Optimists among the manufacturers, as well as independent experts like Robert Williams of Princeton University, assert that a decade of focused research and development could make photovoltaics competitive with newly built fossil fuel power plants.

There is no place where solar energy is always available, and thus a purely solarized world would require energy-storage capacity even where there is plenty of sunlight. Indeed, some regions are so defi-

*Throughout this book, we present costs in 1989 dollars, unless otherwise indicated in the accompanying notes. However, historical references to costs, such as President Carter's $88 billion synthetic fuels proposal, are left as originally stated.
†ARCO Solar was sold to the West German firm Siemens A.G. in 1989.

cient in "sunfall" (also called insolation) for prolonged periods of time that large-scale transfers of energy would be required in a solarized world. In cases of high population density and/or very low sunlight, the imbalance between need and availability is extreme. For instance, during midwinter in New York City, energy consumption slightly exceeds total available sunlight, even with 100-percent conversion efficiency. Of course, the city's fossil-fuel deficit is even worse; and just as the city will never be food-independent, it could never be energy-independent either. The same holds true for much of the densely populated area of the industrial world lying in the high latitudes, where sunlight falls less directly than in the tropics and cloudy days are common.

By the same token, some parts of the world have large sunfall surpluses. Much as food or fossil fuels are transferred from surplus to deficit areas, solar energy could also be shifted and stored in some appropriate medium. In terms of the Earth's surface, not all that much land would be needed to capture the needed sunlight. For example, only 130,000 square miles, an area slightly larger than the state of Nevada,* would be sufficient to meet present-day fossil fuel needs if sunlight were utilized with 10-percent efficiency; and this value is already attainable with today's most efficient photovoltaic modules, even after adjustment for conversion, storage, and transmission losses.

An area fivefold this size could be assembled easily in desert or scrub regions around the world. For comparison, it would amount to only a tiny fraction of the area used to satisfy world food requirements, so even with anticipated growth in demand for energy, land would not be a limiting factor. The main issue would be the willingness of solar-surplus regions of the world to tolerate being used for energy production to supply deficit regions. But the use of land for solar energy is far more benign than that required for other forms of energy exploitation, such as strip-mining or offshore oil drilling. Arid and unproductive land with low value for other uses could be exploited. In the United States, the Nevada nuclear-weapons test

*We are not, of course, proposing to cover a single region like Nevada. Rather, the area would be assembled in small parcels around the world, thus avoiding the environmental impacts of a huge installation.

facility might be a prime candidate for adaptation to this purpose, and solar-energy production would become a major export industry for other high-sunfall areas like Egypt, Saudi Arabia, Mexico, and India, some of which now produce oil.

The Hydrogen (H₂) Connection

The Hydrogen (H_2) Connection

But how can we make solar energy transportable and adapt it for high-power use? The most convenient solution is to convert solar energy to chemical energy and store it in the hydrogen molecule, which can be viewed as a hydrocarbon without the carbon.

Like natural gas, hydrogen can be transported through pipelines and burned to power cooking stoves or to heat homes. It can be stored as a gas, as a liquid at minus 423 degrees Fahrenheit, or as a solid, a chemical compound called a hydride. Any of these forms can be used in place of gasoline to power an automobile. Hydrogen also reacts with oxygen to produce an electric current in a chamber called a fuel cell. This device is like a giant battery; chemical energy stored in the gas is released as a current when it reacts to form water. The current is extracted from the chamber by electrodes, like the poles of a battery. Fuel cells are cheap to build, pollution free, and much more efficient than burners as a device for extracting energy from hydrogen. Coincidentally, fuel cells, like photovoltaic cells, date to 1839, and later found their first practical application as power sources for the U.S. space program. In 1985, a Department of Energy Committee said that hydrogen fuel cells would be commercially feasible within a decade, and a French firm, Alsthom, is now marketing them. They are not widely used because making hydrogen remains more expensive than burning coal.

Today, hydrogen made largely from natural gas or coal is used to manufacture fertilizer, cooking oils, and petrochemicals. Before World War II, a fifty-fifty mixture of hydrogen and carbon monoxide, made by passing steam over hot coal, was widely used in homes in the United States and Europe. It was variously called water gas, town gas, coal gas, or producer gas. But after the war domestic use of hydrogen was supplanted by natural gas. By then, hydrogen was stuck with a nasty image: the explosion of the dirigible *Hindenburg* at Lakewood, New Jersey, in 1937, its frame dancing in the flames like a skeleton jangling in hell. Nearly fifty years later, the *Challenger* space shuttle disappeared as a puff of smoke over Cape Canaveral, the result of a leak in the booster rocket O-rings that detonated the main hydrogen fuel tank.

Yet these accidents should be a warning, rather than a barrier to the widespread acceptance of hydrogen in the future. To be sure, a car fueled by liquid hydrogen is a potential firebomb, but so is a gasoline-fueled car. Tank-stored hydrogen can blow sky-high; but so can natural gas, and it often has. However dramatic hydrogen explosions can be, they are in fact rare. The issue is one of relative safety and the cost of achieving a rough equality of risk with other energy systems. Although hydrogen ignites more easily than gasoline or gas, it is less dangerous in other respects. For example, a hydrogen flame is smokeless, and the gas dissipates rapidly once released. The universal application of hydrogen will require a modest degree of technical advance in safety. But based on several decades of generally successful space-program experience and crash-testing of liquid-hydrogen vehicles, it appears that the overall risks of hydrogen can be made comparable to those of other fuels.

Another point about hydrogen is that although it contains no sulfur or carbon and so yields no sulfur dioxide or carbon dioxide, its combustion does form nitrogen oxides, because airborne nitrogen burns in a hydrogen flame. This is not so much a problem for cars, in which hydrogen-fueled engines can be designed so as to reduce nitrogen-oxide emissions far more effectively than the tailpipe controls now used; but it potentially is for burners of the type found in power plants. The solution might be to extract hydrogen's energy at low temperatures with catalytic burners or with fuel cells.

There are no large reservoirs of hydrogen under the Earth, so we cannot drill for it. But water, or H_2O, contains an essentially limitless store of hydrogen, which could be extracted by a two-step process involving solar energy (a suggestion made nearly three decades ago by John O'M. Bockris, now a professor at Texas A & M).*

The first step is to set up an electrical current with photovoltaic

*It should be understood that the virtues of hydrogen could be exploited in conjunction with any other available nonfossil primary energy source, such as a nuclear-power station, in the place of solar power.

cells. The current is then passed between two electrodes immersed in water, a process called electrolysis, which splits the water molecules into their gaseous elements, hydrogen and oxygen. Energy that started out as sunlight and was converted to electrical energy by solar cells now resides in the chemical bond of the hydrogen molecule. Then we can release the energy as heat by burning the hydrogen gas, or as an electrical current by combining it with oxygen in a fuel cell.

Other storage and carrier options besides hydrogen are available to use with solar or nuclear energy. A totally man-made form of hydropower called "pumped storage" is already used widely, and it could possibly fit with some solar power plants. In a pumped-storage system, water is raised from a low-lying reservoir to a higher one when electricity is available (such as during daylight from solar energy) and then run downhill through a turbine to recover the electricity when the primary source is unavailable. Existing versions are highly efficient, but suitable sites are limited. Moreover, a pumped-storage plant could have severe ecological effects on the body of water from which the water is pumped, and construction of the upper reservoir has the disadvantage of requiring excavation of a hill or mountaintop. Advanced systems will avoid these difficulties by using one artificial reservoir on the surface and another underground (the Japanese are building one under the sea bed).

Batteries are another possibility, but they have limited capacity and durability. In the future, improved versions may be available to store either solar or nuclear power for use in cars. Advances in superconductivity could permit transmission of pump-stored solar power over long distances. But neither batteries nor pumped storage promises a comprehensive solution to transporting energy; only hydrogen can deliver both the power density and the flexibility of fossil fuels.

NUCLEAR POWER

Nuclear-power advocates were quick to take advantage of the growing greenhouse concern in the summer of 1988. "Not So Bad After All?" queried a *Newsweek* headline, and there were moves on Capitol Hill to provide research funds for a so-called "safe" nuclear-power demonstration project. Nevertheless, the advent of global warming has not solved the fundamental economic and safety issues posed by nuclear energy, and there is little reason to expect their imminent resolution. Nuclear power fails the initial test for serious consideration: is it now, or does it promise to be in the future, the least expensive energy option available? In the United States at least, the lack of any new orders for nuclear-power plants since 1978 reflects cost escalation and other difficulties that have beset the industry since the accident at Three Mile Island. Some, perhaps, were inevitable by-products of the government and industry pressing an immature technology. Other difficulties were caused by mismanagement and by smug indifference on the part of the electric utilities, reactor manufacturers, and government to reasonable concerns about public safety. The industry has argued that it is a victim of changing public attitudes, and it has a point. But those attitudes were shaped by Three Mile Island, Chernobyl, and various waste-disposal debacles.

Let us set aside whatever negative impressions exist about the potential hazards of nuclear power in this post-Chernobyl era and simply examine the economics involved. The average price of electricity obtained from recently completed nuclear plants is more than thirteen cents per kilowatt-hour. The most expensive nuclear plant, the $6.6 billion Nine Mile Point Unit 2 station on Lake Ontario in upstate New York, produces electricity costing about twenty-three cents per kilowatt-hour. But the cost of power from new coal-burning plants is only six to seven cents per kilowatt-hour, and a variety of small-scale options like small hydropower dams have been supplying new electricity below six cents per kilowatt-hour.

Coal plants and hydropower dams themselves bear environmental consequences, but nuclear power has one other competitor that has none of these drawbacks: energy-efficiency improvement—in other words, extracting extra useful energy without using more fuel.

Energy is "conservative." When it is altered from one form to another, none is lost (a principle known as the First Law of Thermodynamics). In other words, the same amount of energy can flow through a series of transformations, for instance, from electrical to chemical, as when a car battery is charged. This property allows it to be transformed usefully. But heat is always one of the forms to which it is transformed during a conversion, no matter what the end product desired (the Second Law of Thermodynamics). Energy can never be converted with 100 percent efficiency from one form to another non-heat form. Some heat always flows during a conversion.

With effort, this heat flow may be minimized and efficiency improved. For example, only about one-third of the combustion energy produced in a large coal-burning electric power plant drives the steam generator, while two-thirds is lost as waste heat from the smokestacks and cooling tower. When fuel is expensive, however, enlightened utility officials recapture some of the waste heat to heat homes and drive the engines of industry. Conversely, electricity can be generated as a side benefit of the waste heat from factory engines. Known as cogeneration, this method for stripping useful energy from waste heat is widespread in Europe, where energy prices are higher than they are in the United States. For example, the town of Vesterås, Sweden, simultaneously satisfies all of its electricity and its space heating needs via cogeneration. Unfortunately, in the United States this practice is less common.

Efficiency can be improved not only in devices that generate energy but in ones that use it, too. An efficient lightbulb channels more of its electrical power to lighting and less to heating the sur-rounding air. An efficient refrigerator uses more power to turn its compressor and cool food and less to heat your kitchen. Conse-quently, the same amount of light or refrigeration is delivered using less electricity and, ultimately, less fossil fuel.* Implementing such measures is cheap, with many costing less than two cents per kilo-watt-hour saved, and, as signs in India proclaim, "Electricity saved is electricity generated." In this sort of competitive atmosphere, the nuclear option is no bargain at all.

Here is one case in point. As of 1975, Pacific Gas and Electric planned nine new nuclear plants and one coal plant to satisfy the growth in demand anticipated by the early 1990s. But the Environ-mental Defense Fund demonstrated to the California Public Utilities Commission that the company could avoid the need for the new

---

*Not all wasted energy is lost as heat. For example, the efficiency of a car may be improved by making it lighter, and not wasting energy by lugging around unnec-essary weight.

plants and save money at the same time by peddling devices like efficient lightbulbs, which conserve electricity, or by purchasing electricity from industries that cogenerate. The planned construction of nuclear plants shriveled on the vine. Since then, electricity demand in California has been met with cogeneration, efficiency improvements, and solar power rather than new nuclear or coal plants. Subsequently, the Natural Resources Defense Council and the Conservation Law Foundation similarly demonstrated the advantages of efficiency to utilities in the Pacific Northwest and New England.

There is a certain symmetry in the current positions of nuclear and solar energy. If a law were passed that no *new* source of electricity in the United States could use fossil fuels, efficiency measures could easily fill the gap between supply and demand for the next decade, and at a much lower cost than either nuclear or solar energy. But if we were forced to replace all *existing* electricity sources with nonfossil alternatives so that some new power plants had to be built, turning to either solar or nuclear energy would send the cost of electricity skyrocketing. Both nuclear and solar energy are available right now, but neither presents an economically acceptable alternative for widespread use. Neither will become an option until the price comes down or the public becomes so concerned about global warming that it will tolerate considerable financial sacrifice. Furthermore, both are in the same position with regard to transportation needs: until a cheap storage and carrier medium is available, neither may be used. But here the similarity ends.

THE CROSSING POINT

The distinction between solar power and nuclear power leaps out at anyone who compares the histories of the two technologies. In brief, the price of nuclear power has been soaring while the price of solar power has been dropping rapidly. In 1954, it would have cost $400,000 to plaster a roof with enough of the first Bell Labs cells to power a house. Today, the cost would be only a few thousand dollars. In vivid contrast, between 1971 and 1985, the average construction cost of U.S. nuclear power plants multiplied five times.

The cost of photovoltaic modules has dropped by the same factor since only 1979. Looking at the near future, Chronar Corporation has offered to construct a fifty-megawatt photovoltaic power plant near Palmdale, California, by 1992, which would provide electricity at a lower price than the most expensive nuclear plant, Nine Mile Point Unit 2. The Luz International solar thermal plant already undersells power from the average nuclear plant and is competitive with other sources of peak power. The cost curves have crossed.*

The latest version of single-crystal cells now achieves 22-percent efficiency in the laboratory. With two stacked together and the sun focused with a light-concentrating lens, an efficiency greater than 30 percent has been achieved, and a value of 40 percent seems within reach. Commercially available varieties achieve about 15 percent. Amorphous silicon cells attain 6-percent efficiency in commercial use and 12-percent in the laboratory, but they may ultimately reach 18 percent. Coupled with improvements in production techniques that would follow from mass orders, the efficiency gains will drive the price down rapidly over the next decade.

Based on plausible, if somewhat optimistic, projections of technical progress on amorphous silicon cells, Joan Ogden and Robert Williams at Princeton have described a solar-hydrogen system that could be in place and generating baseload electricity at only 20- to 50-percent higher cost than new coal-fired plants by the end of this

*The construction costs for U.S. nuclear-power plants completed in the mid-1980s averaged $2,500 per kilowatt, *exactly the same cost* that the Chronar Corporation anticipates meeting with its planned photovoltaic plant. The construction cost of Nine Mile Point Unit 2 was $6,100 per kilowatt. Strictly speaking, construction costs for solar and nuclear power cannot be compared directly, for a nuclear plant will operate about two-thirds of the time, while a California photovoltaic plant can provide, at most, about one-third of its theoretical peak energy because the sun is in the sky only part of the day (satisfying only daytime peak demand). This fraction, called the capacity factor, varies from place to place depending on latitude and climate. Photovoltaic power can be baseloaded by adding storage capacity, but doing so adds to the price. By contrast, nuclear-plant operation, maintenance, and fuel costs add another 10 to 20 percent above capital requirements, while for photovoltaic plants these costs are negligible. When these factors are considered, it appears that photovoltaic power plants could now be constructed in some places to produce electricity well below the twenty-three cents per kilowatt hour cost at Nine Mile Point.

century. It would use photovoltaic electricity to make hydrogen gas, which could be burned later in a new superefficient turbine to generate electricity. (A fuel cell could also be used.) Its extra cost is comparable to what utilities pay for standard pollution controls. Such a system would produce less expensive electricity than the average nuclear plant, and far cheaper energy than Nine Mile Point does. By that time, photovoltaic power without storage would also provide a cheap source of peak power in sunny places.

In contrast, the future continues to look grim for nuclear power. No permanent disposal arrangement for high-level nuclear waste yet exists, so the ultimate cost remains unknown. No large U.S. nuclear power plant has come to the end of its life, so the costs of safely "decommissioning" one have yet to be determined. The scandal surrounding military-reactor wastes, which broke publicly in 1988, has created an atmosphere in which tough and expensive disposal requirements are inevitable. As these costs are added to electricity bills, the price of nuclear power will doubtless continue to rise.

According to New York energy consultant Charles Komanoff, an astounding $100 billion has gone into nuclear power plants that were started but never completed and into cost overruns that ended up making plants noncompetitive with other power. Imagine if $100 billion had been made available for research on solar energy! Part of the reason for this was the long planning and construction time of nuclear plants, during which interest rates went out of sight, construction costs escalated, electricity demand tumbled, and a partial meltdown occurred at Three Mile Island, leading to calls for costly safeguards. In contrast, ten megawatts of photovoltaic power, available within two years, can gradually grow to fifty or a hundred megawatts, simply by adding modules on adjacent land. In the meantime, electricity is continuously generated, even if market conditions change. But abandoned nuclear plants never generate a watt, though some have been converted to gas or coal combustion.

Perhaps even more disturbing than its economic problems, large-scale nuclear power is closely related to weapons production. Plutonium, a fuel for nuclear bombs, is a necessary by-product of the fission process that creates nuclear power. A typical U.S. reactor generates about fifteen bombs worth of plutonium every year. The present world nuclear-generating capacity is equivalent to 300 standard U.S. plants.

If all of the world's energy were to come from nuclear power, at least 12,000 such plants would be required. Keeping the plutonium wastes out of the hands of those who would use them for terror or war would require entirely novel global security arrangements.

With the price of solar power becoming comparable to that of nuclear power, and still falling, one wonders why nuclear energy has any enthusiasts at all. The love of the big and the complex is one possible reason; certain military exigencies are another, such as the desire to use access to materials from which bomb fuel could be made by other countries as a political tool, and the need for backup reactors that could be adapted for weapons production in emergencies.

It would be a mistake, however, to deny that solar power has its own problems or to assert that nuclear power is totally devoid of hope. With solar cells made of silicon, manufacturing waste and disposal problems are minimal. But nonsilicon cells are now under development that could cause serious difficulties if mishandled. Amorphous thin-film cells made of copper indium diselenide or cadmium telluride present both occupational and environmental problems because they are composed of toxic metals. Waste minimization and recycling must be planned in advance of large-scale production or photovoltaics may run into problems reminiscent of nuclear power.

In parts of the northern United States or northern Europe, where power demand can peak during winter, local solar power would be at least three times as costly as that in southern locales because the same modules would provide much less energy, and the mismatch of peak demand and peak supply would require additional investments in backup fuels. Solar power will penetrate the market gradually in optimum sun areas, but full exploitation will require the hydrogen connection. In a limited way, that connection may not be far off: a German firm is already examining the use of cheap Canadian hydropower to split water for hydrogen that can be shipped to Europe for industrial uses. One day, solar power will be the source, and a desert country will be the exporter.*

*Land requirements for hydrogen production may be somewhat greater in the driest of desert countries than appears from calculating the area needed for photovoltaic modules. Inasmuch as freshwater would be in short supply, space would also be required for solar desalinization of seawater before electrolysis.

Meanwhile, the nuclear option will linger on, not alive but not quite dead, beset by continuing failures, the lack of permanent waste disposal, the potential for proliferation of weapons-grade fuel, and excessive costs. The theoretical possibility of solving these problems will continually feed hopes in some circles; particular attention has focused on "inherently safe" reactor designs, which supposedly would not cause accidental radiation dispersal, even if operated by an idiot. Development of a cost-competitive reactor of this type is a prerequisite for public acceptance of nuclear power. No affordable, safe reactor exists now, though the High Temperature Gas-Cooled (HTGC) model had been seen as a step in this direction. This reactor's glass-enclosed fuel pellets are designed so that they will not get hot enough to melt under foreseeable circumstances. Yet only a month after a *New York Times* story touted the advantages of this design, the shutdown of the only commercial-scale model in the U.S., a 330-megawatt reactor at Ft. Vrain, Colorado, was announced; operating costs had been astronomical. Early in 1989, the West German government started backing away from its commitment to continue operations of a 300-megawatt HTGC reactor that had been shut the previous October because of safety concerns.

Other proposals, fueled largely by the federal treasury, will sporadically be presented in attempts to revive the nuclear industry from its comatose state, but several considerations make nuclear power a poor bet as a major global source of energy in the future, regardless of technological fixes. After Chernobyl, public attitudes about nuclear power have turned so skeptical in Western Europe, the East Bloc, and the United States that any new large commitment to commercial-scale experimentation is out of the question for the foreseeable future. If the global public were forced to choose between nuclear power and global warming, they would probably choose to suffer global warming, at least until its painful consequences were clearly manifest. As of 1989, European politicians were running for cover on the issue. Following a referendum in Sweden, the government plans to retire its entire nuclear capacity, which provides half its present electricity, by the year 2010, and Italy is doing the same. In West Germany, the once pronuclear Social Democrats have turned 180 degrees, as has the British Labour Party; and in the Soviet Union public demonstrations have at least temporarily halted some reactor projects. It may be difficult for Americans to comprehend the intensity of anger, fear, and disgust engendered in Europe by the fallout from Chernobyl, which necessitated the destruction of crops, milk, and animals; but clearly the tide has turned on attitudes about nuclear power.

Even if public opinion were more favorable, however, inexpensive experimentation with new reactor designs is impossible because catastrophic failure, for all its devastating potential, does not occur frequently. As a result, many reactors must operate in the real world for years before a reliable estimate of the likelihood of various accidents can be made. Thus, the technology must go to commercial scale to determine whether it is safe enough to be allowed to do so! By contrast, solar technologies are either already tested or can be tested on a small scale; and, in any case, they do not entail the potentially catastrophic consequences of nuclear failures.

To be sure, there are examples of relatively successful nuclear programs. France has constructed some reactors at competitive cost, in the $1,000-per-kilowatt range, and generates 70 percent of its electricity from nuclear power. South Korea and Taiwan are proceeding with large-scale commitments; these countries are able to take the financial risk partly because technological issues have not yet become a matter of public debate. But when one considers the waste and proliferation problems, and the cheap availability of energy efficiency as an alternative, nuclear power has poor prospects, either in the short or in the long term.

THE SOLAR MOSAIC

Besides photovoltaics and solar thermal electric generators, there is a veritable cornucopia of alternatives for capturing the sun's energy, each of which is practical somewhere now. Just as thermal electric plants use arrays of reflectors to trap the sun's heat in oil, artificial solar ponds trap energy in a hot layer of water. But nature itself converts sunlight into heat and other forms of energy, which can be trapped and utilized. Much as hydropower snares some of the solar energy that drives the atmosphere's evaporation-precipitation cycle, wind power catches some of the sun's heat that keeps the air in motion, and electricity can also be generated from heat trapped in the ocean. Tidal-energy dams* indirectly utilize the gravitational forces tying the Earth, sun, and moon together, while geothermal systems tap remnant energy from Earth's formation. Some of these systems, like rooftop water heaters, may be constructed on a household scale; others could provide energy community-wide.

Biomass energy—solar power derived from once-living organisms—turns the deforestation problem on its head. Cattle dung, crop residue, firewood, gas or methanol distilled from wood, and ethanol fermented from corn or sugarcane can all be burned to release solar energy stored originally by plants. This process also releases carbon from the plants' organic matter in the form of carbon dioxide. But if as much wood, or corn, or crop residue, or cattle fodder is regrown as is burned, then an equal amount of carbon dioxide is removed from the air by photosynthesis. The new growth can then be used as fuel while producing no net carbon dioxide if an equal amount of crop is again cultivated. If energy needs grow, cropland can be expanded. As long as harvested fuel equals new crop, there is a balance.

This sort of energy is called renewable (a term applied to all solar energy, in fact) because it can be continually drawn on as long as the sun shines, unlike coal or oil. The trouble with deforestation, of course, is that trees are cut and burned and less biomass, if any, is grown afterward. Indeed, as we note in chapter 5, deforestation could in theory be reversed by *reforestation* on a large scale, which would absorb significant amounts of carbon dioxide emitted by fossil fuels. The wood from reforested areas could then be harvested for renewable energy if equivalent replanting occurred.

About 15 percent of global energy comes from biomass; in rural Asia biomass accounts for 80 percent of the energy, though it is generally utilized by very inefficient burning of wood, crop residues, and dung. But advanced forms of biomass energy have substantial potential in the Third World in particular. For example, biomass can

---

*Tidal-energy dams, such as the one proposed for the Minas Basin in the Bay of Fundy, could have severe environmental consequences.

be converted to methane† which may be burned at extremely high efficiency in stoves and turbines to provide much more energy than traditional methods. Brazil fueled almost 20 percent of the country's 12 million cars in 1988 and 90 percent of the new cars with ethanol produced by fermentation of surplus sugarcane. The program, which receives a hefty government subsidy, was developed in order to cut oil imports, not for environmental reasons. Indeed, supplanting gasoline completely would eventually add to the pressure for land, which is contributing to deforestation. But Brazil is slowing its commitment to ethanol because it cannot afford the subsidy needed to make it competitive with low world oil prices.

Theoretically, the United States could supply a substantial portion of motor-vehicle fuel from ethanol derived from corn. In reality, the energy intensity of U.S. agriculture is such that it may require more energy to produce ethanol than can be extracted from it as a fuel. This would make ethanol expensive once expanded production exhausted surplus crops. If this production energy came from fossil fuels, ethanol would lose its greenhouse benefit entirely. Moreover, large-scale use of ethanol would not be problem free; its combustion produces emissions that cause acid rain and smog, though less so than gasoline. The United States has additional biomass potential. For example, some U.S. dairies are already fully powered by methane generated from the manure their cows produce.

Projections indicate that renewable biomass could supply 20 percent of global energy demand in the future. These projections may be overly optimistic, for area requirements are larger than for photovoltaics and involve fertile land. In fact, achieving this goal would mean utilizing 10 to 20 percent of the current cropland on Earth. Along with other forms of solar energy, however, renewable biomass could be part of the total energy mosaic for the twenty-first century.

†There are two types of processes for producing methane. Digestion relies on bacterial decomposition (of dung, for example), while gasification by distillation of wood or crop residue is a chemical process analogous to synfuel production.

THE BRIDGE

In a solar-hydrogen world, carbon dioxide, acid rain, smog, and oil spills could be eliminated. Yet industry, transportation, and habitation cannot be altered immediately, even once all of the technical issues are resolved and even if the cost of change proves to be modest. Optimists believe that ten years will pass before photovoltaics can seriously compete with fossil-fuel electricity sources at a reasonable price, by which time global warming should have become obvious. At that point, as we mentioned in the last chapter, action could still be slow in coming. Governments might take another decade to provide the full range of incentives and regulations needed to bring about a solar-hydrogen economy (see chapter 9), and, as experience in other industries has shown, it could take another three decades to gear up production, establish an infrastructure, and fully exploit the technologies. Resolution of some technical issues, particularly those surrounding the hydrogen car, could take longer still. At this pace, at least twenty years would pass before conversion to a solar-hydrogen economy was begun in earnest, and perhaps fifty years or more before it was completed. But global warming won't wait that long. With the greenhouse effect intensifying, we need a bridge to close the gap between the here-and-now and the fully solar world. That bridge will be energy efficiency.

Efficiency has one unbeatable characteristic: burning less fossil fuel to get the same amount of energy means proportionately less carbon dioxide produced. The potential for efficiency that was unearthed in California only scratched the surface. Other opportunities abound. Toyota's prototype AXV diesel car averages more than 110 miles per gallon on the highway, using about one-quarter of the fuel needed by one of today's twenty-eight-miles-per-gallon cars. As a result, it emits about one-quarter of the carbon dioxide. Because energy comes at a price, there is often an economic benefit to efficiency; in this case, although the initial price of an efficient car is generally higher, even more is saved over time on fuel costs. Unfortunately, the high price tag discourages buyers regardless of eventual savings.

Energy-efficient refrigerators now available use less than half as much electricity as do standard models and keep food just as cold.

Because their electric power is usually generated from coal, oil, or natural gas, the efficient refrigerator produces half as much carbon dioxide at the power-plant end of the line. At the technological cutting edge, a small California company, SunFrost, produces a refrigerator that cuts energy use by 85 percent.

In addition to lower carbon-dioxide emissions, reduced electricity use also translates into proportionately less nitrogen oxide and sulfur dioxide if fossil fuels are the source of power. In the case of automobiles, the relationships are more complicated because combustion conditions may be different in the different types of cars, so the relation to amount of fuel burned is not so simple for the pollutants other than carbon dioxide. But many new design features can simultaneously increase fuel economy and reduce emissions of air pollutants. Less air pollution will generally mean slower growth of other greenhouse gases, methane, and tropospheric ozone. So we count five direct benefits from energy efficiency—less carbon dioxide, carbon monoxide, sulfur dioxide, nitrogen oxides, and hydrocarbons—which in turn produce three indirect benefits—less acid rain, less tropospheric ozone, and less methane.

Nearly every product or process, from the generators that convert fossil fuels into electricity, to the motors in factories that turn out products, to the products themselves, can be designed to use less energy and to produce less carbon dioxide. Unfortunately, few things are designed specifically to maximize energy efficiency because energy consumption is just one factor among dozens that are customarily considered in product or process design. Sometimes efficiency means a straightforward tapping of waste heat, as in cogeneration. In other cases, the old product must be thrown out and one of an entirely new design substituted.

Consider the electric lightbulb. When the forerunner of the incandescent bulb, which converts electricity to light by causing a filament to glow, was invented in the 1840s, there were two major obstacles to production. First, the filament had to remain solid, and not melt, at the high temperatures required to make it glow. Second, a means of housing the filament in a vacuum had to be developed, for the high temperatures would result in rapid oxidation and disintegration of the filament in air. Considerations of energy efficiency were beside the point, because electrical service wasn't widely available in 1879 when Edison solved these problems with a carbon filament in an evacuated glass envelope. The incandescent bulb reached its current form before World War I with the introduction of the tungsten filament, and the design has stagnated ever since. Propeller planes evolved into jets, vacuum-tube radios into solid-state models, but the incandescent bulb has endured; and the original disregard for energy efficiency remains locked into its design.

After World War II, a considerably more efficient light source, the fluorescent tube, became common in many factories, offices, and stores. Instead of converting a current into light by passing it through a filament, a fluorescent tube illuminates by discharging a current through a gas, causing the tube to glow. As the Second Law of Thermodynamics dictates, the light from an incandescent filament is inevitably associated with heat; both are produced simultaneously when electrons in the current collide with the atoms that compose the filament. In fact, incandescent bulbs waste 95 percent of the electricity they draw, turning only 5 percent to light. Heat loss from fluorescent lights occurs too, but it is minimal, so they are much more energy efficient.

Fluorescent tubes also have longer lives than incandescent bulbs, and when their lower energy use is factored in, they are cheaper in the long run. But incandescent bulbs are still widespread in residential use for three reasons: lower initial outlay by the consumer, the gentler light they cast, and their compatibility with existing fixtures.

Now, new compact fluorescent bulb-shaped lights are becoming available that can compete with incandescent bulbs in terms of their light quality and socket compatibility. The new compact 18-watt fluorescent bulbs give light equal to that of a 75-watt incandescent bulb, and they use only one-fourth of the electricity and last more than ten times as long. They cost about $15 apiece, but each one saves $20 to $30 in electricity bills over its lifetime. Still, the high price has been a major obstacle to its acceptance, and thus the bulbs are almost entirely absent in homes. But there are other options for those who find it hard to swallow this initial expense. Money and energy savings can be squeezed from the Sylvania "Capsylite," an incandescent tungsten-halogen bulb that costs between $2.50 and $4.00, uses just two-thirds the energy of a standard bulb, and lasts fully four times as long.

Aside from the incandescent bulb, hundreds of sources of wasted energy hide like mice in the walls of the global economy, and as long as they do, money and energy get eaten away around the clock. Like inefficient appliances, some of the mice remain hidden for years while others, like lightbulbs, can be quickly trapped and eliminated.

Why is so much energy wasted? Either the more efficient products are, like the new fluorescent bulb, cheaper to use but so costly to buy that the so-called long-term savings are irrelevant to the consumer. Sometimes the individual who pays for the product, for example, a landlord, is different from the one who will pay for the energy, as is usually the case in rental housing. Sometimes there is a lack of information and understanding on the part of the consumer, or there is consumer reluctance to try new products.

How much wasted energy exists in this form? In other words, how much could be saved at a net economic gain with no sacrifice of product or service quality? Estimates vary widely, from those made by Amory B. Lovins of the Rocky Mountain Institute, who says that global demand could be cut by one-third over the next two decades and eventually by at least 75 percent, to the Monsanto executive who commented that "if industry thought it was cheaper to be energy efficient, it would have already done so," a view frequently echoed by economists.

Here we shall stake out a moderate position, between the pessimists and the optimists. Business executives make misjudgments (just look at Detroit, when the West Germans and then the Japanese began churning out smaller cars); consumers sometimes act unwisely; and thus not all parts of the energy-supply system act perfectly (particularly because much energy supply and use is regulated by the government in a manner often guaranteed to lead to waste). We do not know how far we could get with efficiency and still save money, but surely much energy is out there to be saved at a profit.

Moreover, recent events have sharply undercut the argument of the pessimists. When energy prices rose sharply in the 1970s, energy use stopped growing, its higher price stimulating the search for savings and substitution. The U.S. Gross National Product grew 40 percent between 1973 and 1986, yet energy use hardly budged. Instead, the economy became much more energy efficient. By itself, this statistic doesn't prove that other energy savings remain for the asking, but the experience reveals that governmental efforts, in particular, pay back handsomely. Greater progress was made in residential energy efficiency in California than anywhere else because the state's regulatory agencies mandated innovative programs like the ones mentioned earlier. In one case, Southern California Edison gave away 450,000 compact fluorescent lightbulbs to low-income rate payers. The utilities themselves profited nicely because efficiency enabled them to avoid investing in new, costly power plants.

We also know that other countries, albeit with higher energy prices, have developed vastly more efficient economies. Japan and West Germany both produce a unit of gross national product with only half the energy expended in the United States. Some of this difference is attributable to life-style choices that Americans would deem unacceptable, such as the small, cold homes the Japanese endure. Yet Japanese refrigerators are more than twice as efficient as current U.S. models, and their cooking stoves use only one-quarter of the energy that American stoves do. Logic dictates that further energy savings could be realized here if only we would devise proper incentives. The same applies even more to the centrally planned economies of the Soviet Union and Eastern Europe, which are far less efficient than the United States.*

Building the Bridge

Fossil-fuel energy not used equals carbon dioxide not emitted—it is that simple. Over eighty percent of global energy comes from fossil fuels, and a big bite out of energy use will take a big bite out of carbon dioxide. Climate change probably won't be halted in this way, but the warming can be slowed significantly. One realistic estimate provided by the Stockholm Environmental Institute is that energy use can be kept constant between now and 2025 through efficiency improvements with the global economy still expanding at a healthy clip. Otherwise, the world could be committed to a six- or seven-degree warming by the same date.

---

*The restructuring underway in the East Bloc provides a unique opportunity to construct energy-efficient economies and simultaneously reduce the high level of air pollution there.

What will slower warming achieve? It will buy us that most elusive commodity, time. As Henry Ford said, "Time loves to be wasted. From that waste there can be no salvage. It is the hardest of all waste to correct because it does not litter the floor." We will have more time to put the photovoltaic-hydrogen system in place before disaster strikes and, in the interim, climate change will slow to a modest rate that society and nature can accommodate.

Most significantly, energy efficiency and environmental benefits go hand in hand with improved, not reduced, economic performance. Fifteen percent of the U.S. investments in new plants and equipment in the late 1970s went into the electric utility sector. The same electricity needs could have been handled at less cost with efficiency improvements, leaving billions of dollars for other productive investments. Although the utility share ballooned over the period thanks to the highly wasteful investment in nuclear power, a smaller drainage of capital occurs continuously. For U.S. industries, the cumulative inefficiencies in production processes contribute at least marginally to the decline in competitiveness with the Japanese and other more efficient economies. For comparison, the United States now spends 11 percent of its Gross National Product to obtain energy, while Japan uses 5 percent.

Once implemented, efficiency improvements will continue to pay off even as the new energy sources are installed. Over time, photovoltaics, solar thermal generation, biomass energy, and wind power will all play a part. Residents of a new superinsulated house in the northeastern United States will stay warm all winter using only one-sixteenth of the heating fuel required for a standard house. Low carbon-emission natural gas will provide the minimal heat needed, while a utility will provide electricity using advanced gas turbines with nearly twice the efficiency of the old coal boilers.

For the long run, the photovoltaic-hydrogen system provides a plausible alternative to fossil fuels. It can accommodate a society reliant on cars or on mass transit; it can satisfy energy needs in both rural communities and dense urban areas. Other options exist, but the virtue of photovoltaic-hydrogen is that it is constructed from clean and inexhaustible sources. Furthermore, it is modular. If a better large-scale primary source reveals itself, photovoltaics can still play a small role.* Similarly, if hydrogen works best for only some uses, a different storage medium, such as batteries, might suit others (for instance, if electric cars become popular).

We do not need to take another one-shot gamble, as the utility industry did with nuclear power. In fact, the 1989 flap over cold fusion power smacked of the old nuclear-style desire for a high-energy quick fix. Had cold fusion turned out to be a reality, it would still have required many decades of basic research. The same is true for its close relative, hot fusion. Photovoltaics, by contrast, are already here, in commercial quantity.

Quite soon, photovoltaics alone will provide power at peak demand hours at optimum locations; later, photovoltaic-hydrogen systems will do so elsewhere. Liquid hydrogen may first power airplanes, as it now fuels rockets; then locomotives will be powered by fuel cells, and fleet vehicles such as city buses will be fueled with hydrogen at central stations. Hydrogen will be mixed with natural gas for industrial power, then used on its own. Part by part, piece by piece, the current energy pattern can be replaced like chips in a mosaic until a totally new picture emerges. A few pieces are added, a few taken away; one day it will all have changed, but nothing will have happened overnight.

*Some experts are touting a "wet" photovoltaic system, which produces hydrogen directly without the intermediate transmission of electricity.

CHAPTER 7

Sic Transit:

The Rise and Fall of the Automobile

THE ENERGY-INEFFICIENCY MOUSE EMBEDDED IN THE AUTOMOBILE
is as big as a rat and as old as the Model T. It has been devouring gasoline since 1901, when oil began gushing from Texas's huge Spindle Top field at a nickel a barrel. The mouse was conceived in the mating of two desires: fast, high-power driving and easy, single-clutch shifting. As auto historian James Flink puts it, "Compared with Europe, . . . the seemingly inexhaustible supply of cheap gasoline and lack of any horsepower tax allowed the American manufacturer to sacrifice engine efficiency to obtain maximal engine flexibility. The characteristic large-bore, short-stroke 'American Engine' that emerged permitted the American driver to reduce shifting to an absolute minimum."

The craze for high-performance vehicles was already in high gear by 1906, when *Scientific American* noted the move to "a steady increase in cylinder capacity, which is due, no doubt, to the national temperament which makes the driver prefer, if possible, to run continuously on the high speed, even when sharp hills have to be negotiated." Or, as *World's Work* magazine observed at the time, "Nothing is more unsatisfactory than a car too heavy for its horsepower." Evidence of interest in fuel economy by early consumers is slender; indeed, a perusal of early car advertisements finds no mention of it. Interest in mileage centered on driving range. Even so, some early models were paragons of efficiency in comparison to what came later. The 1906 Model N achieved twenty miles per gallon, half again as much as the average mileage of cars before the enactment of U.S. fuel-economy standards in 1975. The cause of the decline in mileage was the continual oversizing of the engine: compare the 20-horsepower, 1,200-pound Model T to the 150-horsepower, 4,337-pound beast of a 1974 Buick Le Sabre.

The inefficiency that still plagues automobiles is a major contributor to global warming because motor vehicles consume about one-quarter of the energy used by industrialized nations like the United States, and provide a like factor of their carbon-dioxide emissions. The impending growth of car ownership in the Third World, should it be based on current-generation gasoline models, would unleash a torrent of emissions over the next few decades, inexorably accelerating the greenhouse effect. The difficulty of reversing this pattern, once it has been set, is apparent when we consider that the automobile has gained a stranglehold on the United States in spite of the unprecedented environmental costs it imposes. The reason is not just that people love cars, but that cars have largely defined the evolution of economy and culture in twentieth-century America.

When visionaries speak of replacing the automobile, their considerations go well beyond technology. Their analysis is necessarily also a critique of modern culture. To quote Richard Register, an urban ecologist in Berkeley, California, "Without question the most destructive agent of social disintegration, ecological contamination, poisoning of people and environment, waste of energy, and even homicide (outstripping violent crime by more than two to one) is the automobile." Gasoline automobiles are at the very center of a web of social and economic relations. They cannot merely be plucked out and replaced without fracturing the whole structure. Nevertheless, we can alter this web gradually, one segment at a time, much as the energy-delivery system can be altered.

To understand the elements of such a reorganization, consider first what people want from the automobile, be it unfettered mobility, speed of travel, or an expression of personality; and second, how to deliver these features but with smaller climate consequences, through technological improvements such as higher mileage efficiency and use of new fuels. Third, we must explore changes in the socioeconomic web like urban redesign, which could reduce overall dependence on automobiles, thereby cutting emissions while also reducing traffic congestion and urban sprawl. Such measures go to the heart of restructuring, and to many people they are as important as limiting the greenhouse effect. A reasonable argument can be made that efforts short of total restructuring are ultimately wasteful and duplicative. But devolution of the car culture will not occur overnight, while the onset of warming is already here. We cannot afford to wait for the Messiah.

FATEFUL CHOICES

To the extent that people give any thought to the question of how we got where we are today, they probably think the direction taken was a matter of course. In fact, individually owned gasoline-powered automobiles beat out several other transportation options, following the expression of specific preferences over other alternatives. These were sometimes individual choices, sometimes governmental, sometimes corporate. The choices were at times uninformed, stupid, or venal; yet no one should be deluded into thinking that cars were foisted on the public, or that other options were entirely absent. America, in particular, wanted the gasoline car. To understand why this choice was made is to understand the possibilities for change in the future.

In the year 1900, 4,242 automobiles were produced in the United States. The majority were powered by steam or electric batteries, while only about a thousand ran on gasoline. Steam cars originally burned wood or coke, and later, oil to boil water, which drove a piston. The steamers started slowly, ran out of water fast, generated low horsepower for their weight, and froze in winter. But a wide gear range made for easy shifting. Most important in retrospect, boilers could be designed to emit little pollution.

Steam cars descended from the very first self-powered vehicles; Nicholas Joseph Cugnot, a Frenchman, built the first road steam carriage in 1769, the same year that James Watt obtained his original patent on the steam engine. Steam coaches resembling off-track locomotives trundled about the back roads of England in the first few decades of the nineteenth century. Then fear, social conservatism, and engine explosions brought on taxes and safety requirements (such as four-m.p.h. speed limits and an accompanying flagman) that drove them to extinction after 1865. A more contained form of transportation, the railroad, had replaced them.

In the United States, the first practical self-powered vehicle, Thomas Blanchard's pine-fired steam coach, hit the roads of Springfield, Massachusetts in 1825. It didn't catch on either. But when car mania struck in the 1890s, the steam engine again led the way: Ransom E. Olds of Lansing, Michigan and Oldsmobile fame made the first U.S. car sale, a steamer, in 1893; and by 1899, 2,500 steam cars were being produced by thirty different companies. In 1905, a Stanley Steamer reached a record speed of 127.66 m.p.h. at Daytona Beach. But the internal-combustion engine was coming on fast, and its horsepower advantage finally killed steam for good in the 1920s.

Electric cars, made feasible by the spread of commercial power-generation, were seen on the streets of Chicago as early as 1892. Although Midwesterners generally favored the gasoline engine, electric and steam cars dominated the populous Northeastern market at the turn of the century. Electric cars were clean, quiet, and didn't need hand-crank starting, so women preferred them. They were an elegant substitute for horse carriages in cities, where the drawbacks that ultimately proved fatal mattered less. The disadvantages were their short range, fifty to eighty miles per battery charge, and the long recharge time of eight hours. The purchase price was also 50 to 100 percent higher than a gasoline car, and recharge was expensive: about $15 for a New York-to-Boston trip, four times the cost of tanking up with gasoline. Like steamers, electric cars lacked the instantaneous blast of internal combustion, and when Charles Kettering replaced the hand crank with an electric starter on a Cadillac in 1912, the electric car was blown out of the competition.

The internal-combustion engine waited for the car the way that Cinderella waited for the glass slipper. Invented by Belgian Étienne Lenoir in 1860, it seemed destined for use in factories, a junior partner to the steam engine, until 1885 when Gottlieb Daimler and

Wilhem Maybach put one of Nikolaus Otto's spark-ignition models into a prototype vehicle in Germany.* Another German, Karl Benz, started with motor tricycles in 1885 and graduated to a four-wheeler in 1891. In 1889 in Springfield, Massachusetts, brothers Charles and Frank Duryea built the first American gasoline car, and it was Charles, in fact, who won the 1895 race that originally drew Ford's attention to the gasoline engine. During this period the engine was moving from the back of the car, or from underneath it, to the front. This reconfiguration gave more to the car than room: a forward-surging profile began to replace the dowdy horse-carriage look. A new set of symbols heralded an entirely new era and, for better or worse, a new approach toward the pace of life.

Entranced by developments in Europe, Americans so eagerly awaited the car that before any gasoline models had been sold, two motoring magazines had hit the newsstands. The 727-mile Paris–Bordeaux–Paris race of 1895 generated such wide publicity and interest that it was followed by an avalanche of 500 motor-vehicle patents in the United States. Thomas Edison declared that "the horseless vehicle is the coming wonder. . . . it is only a question of time when the carriages and trucks in every large city will run with motors." By 1899, about 1,000 small shops were experimenting with vehicle prototypes, an unprecedented unleashing of skill and invention.

Roy D. Chapin drove a gasoline-powered Oldsmobile from Detroit to New York in 1901 (9 days, 820 miles, 14 m.p.h. average, 27 miles per gallon). Four thousand such Oldsmobiles sold in 1903 just as Henry Ford was setting up shop, and several intrepid souls drove across the continent between 1903 and 1905 (it took about two months), establishing for good the reliability of the technology. The imminent arrival of the cheap automobile, available to the mass market, was widely anticipated, the way the arrival of home computers was foreseen a decade ago. The United States had appropriated a European technology and made the best of it, in the same way as Japan appropriated the video cassette recorder and mass-marketed it in the 1980s.

*George B. Selden of Rochester, New York appears to have had the idea first, but he held back his patent until 1895 and never built one.

Americans looked lovingly at the car in large part because they had become disenchanted with another mode of transport: the horse. The animal had worn out its welcome as the country urbanized. Ironically, the car was viewed as a great advance in public health; Chicago's health commissioner declared that "ideal health conditions cannot prevail so long as horses are in the city," because of the problems associated with manure. Under some circumstances, the horse was a comparative economic burden—keeping a horse in a city could run to $35 a month, twice the cost of maintaining a car—and the capital cost of a horse carriage rivaled that of a horseless one. Horses were slow and unreliable to boot. So when Ford's breakthroughs in the organization of production brought a car's price down below $1,000 and wages up to $5 a day, the car was headed for the mass market.

When we review the gasoline car's advantages over various competitors in terms of convenience, performance, speed, range, cost, reliability, and even cleanliness, it is hardly surprising that it beat out the electric car, the steam car, and the horse. But the gasoline car became much more than simply a mode of transportation, virtually redefining society; and part of that process arose from choices that were perhaps *not* so inevitable. Decisions giving preference to the automobile over mass transportation, instead of creating a balanced system, can be ascribed in part to the public's desire for the peculiar sense of freedom the gasoline car provided and to the necessity for the owner to amortize its initial cost over a wide range of uses.

But officials also sought to limit public expenditures. As people pushed out into then-suburban areas, the extension of subway, elevated, or light rail (trolley) lines to satisfy their needs would have entailed the high initial cost of laying rails with only relatively few riders to share the expense at first. Fare increases to finance expansion of rail service were politically unpopular,* and real-estate developers were pressing for a far-flung commitment to road building. The individual car gave planners an easy way out. With fuel cheap and their emissions not yet a concern, buses also became popular; but rail systems remained at a permanent disadvantage because right-

*Private as well as public operations were subject to public-fare regulation.

of-way purchase and track extension became much more expensive once the area to be served became far-flung and thickly settled.*

Added to the lack of forward vision was hostility to mass transit by some officials and venality by corporations with an interest in motor vehicles. Electric trolleys began replacing horse-drawn coaches as transportation in the 1880s and, by 1900, they had allowed cities like Los Angeles to spread and suburbs to explode along specific corridors. But in 1926, General Motors began the systematic purchase and destruction of trolley lines across the country, and by 1950 it had replaced street cars with its own buses in more than 100 cities.† The company was eventually convicted of criminal conspiracy, but the fine of only $5,000 was, in the words of urban historian Kenneth Jackson, "less than the profit returned from the conversion of a single street car."

Robert Moses, for decades the public-works czar of New York, both city and state, was positively hostile to public transportation. Robert Caro's book *The Power Broker* documents how Moses not only destroyed the existing trolley system but also assured that the expansion of roads left as little room as possible for mass transit. Ignoring the arguments of planners, he built the Whitestone Bridge across the East River in the late 1930s with a light load-bearing structure, thus foreclosing the possibility of its carrying subway lines like two of the older city bridges. At the time, most city families didn't even own cars. And as a result, commuter connections between Long Island and the Bronx, Westchester County, and New England are still restricted to motor vehicle traffic. Moses designed attractive, free-flowing parkways as early freeways, but he purposely built low bridges so that the roads would be impassable for buses.

The trend to more roads and less rail accelerated after World War II. The Interstate Highway Act was passed in 1956, just as railroad

*New York City's Borough of Queens provides an unusual counterexample of the utility of rail service, where rails and development grew outward together from a central business district. According to the Metropolitan Transit Authority, today about two-thirds of commuters from Queens into Manhattan travel by subway, and only about one-seventh by car.
†Although the decline of the trolley had begun a few years earlier, it is perhaps no coincidence that 1926 also saw the peak in ridership for the urban transit industry as a whole.

passenger service was declining. Rail's rebirth in the form of Amtrak was then aborted by federal budget constraints. Over time, the choice of the car was literally setting society in concrete.

Unfortunately, as L. J. White wrote in 1971, on the brink of an era of upheaval in the auto industry, "The strategy of concentrating on model changes has meant an absence of fundamental technological change in the product the industry has produced."* After Ford's production innovations in the first two decades of this century; Alfred P. Sloan, Jr.'s management, marketing, and financing innovations at General Motors in the twenties; and the advent of collective bargaining and Moses' development of the parkway concept in the thirties, the automobile by and large ceased to be a source of new thinking. Even as regulation and foreign competition forced new approaches in the 1970s, technologies were adopted (such as computer controls) that had originated in other industries.

The vast suburban expansion after the war, as the government encouraged mortgage lending, may have seen the apotheosis of a new social order. But it was based on the expression of technological innovations long in place.

END OF THE ROAD

Partly because of natural advantages and partly by invidious design, we find ourselves in a society based largely on a single transportation mode for short, medium, and often long distances as well. One consequence is a continuous expansion of the greenhouse effect. Nevertheless, there are signs of hope. Just as in the case of other

---

*What innovation took place occurred largely in manufacturing technology, not automative technology. Only a few significant developments, like automatic transmissions, were introduced after the 1920s, and between 1945 and 1970 technology hardly changed. Stasis arose, in part, from lack of competition as the 1,000 small shops of 1900 were reduced to three major manufacturers. As a consequence, the U.S. share of the global market fell from over half in 1960 to less than one-quarter over the ensuing twenty-five years, as the U.S. industry responded more slowly than others to the changing conditions brought on by oil-price hikes and fuel-economy and air-pollution regulations.

energy uses, significant improvement in automobile efficiency began with the oil crises of the 1970s. And, as with the development of photovoltaic cells, it may be argued that alternatives to the Otto cycle engine are not far over the horizon in any event. This deterministic view of technology provides little comfort if the process takes a century to unfold. But the existence of a natural evolutionary path at least gives government an existing process to accelerate as it tries to slow global warming.

Being so dependent on one technology makes it hard to envision a world based on another. Yet engines, like fuels, have come and gone over the past two centuries. By 1836, the steam technology used in railway locomotives had evolved to a form that did not change in any fundamental way until the 1890s, much as the automobile remained static for decades until the 1970s. Inefficiency was inherent in both technologies. Compare this comment by railroad historian A. W. Bruce with James Flink's remark at the beginning of this chapter: "The low overall efficiency of the steam locomotive has been tolerated in the U.S. [vs. Europe] because of the simplicity of the locomotive design [and] the low cost of fuel."

Inefficiency is waste energy, and you can actually see it if you know where to look. As we noted earlier, the waste energy of a power plant is carried out the smokestack or cooling tower. The waste energy of a steam locomotive was obvious in the smoke coming from its stack and in the steam swirling around the pistons, both producing hot air instead of motive power. Although locomotive efficiency improved somewhat during the nineteenth century, as steel replaced iron in its manufacture, the need to extract yet more power to lug bigger loads led to the practice of "compounding," similar to cogeneration. First, another cylinder was added to squeeze some of the remaining energy out of the steam after it had accomplished its initial task of pushing out the piston driving the wheels, energy that was otherwise frittered away into the atmosphere. Later, about 1910, the secondary cylinder was succeeded by the superheater, a device that extracted some of the wasted smokestack heat and recycled it to the main cylinders.

Compounding, of course, had its limits, and each successive addition extracted less energy at higher capital cost. Furthermore, steam locomotives were under pressure from antismoke laws constraining their use in urban areas. As a result, compounding was followed by a gradual replacement of steam technology altogether. Beginning about the turn of the century, a totally electric locomotive slowly penetrated the railway system, first as a switching engine in yards and then in urban terminals in Chicago and the new warrens built beneath Grand Central Station in New York. The diesel-electric locomotive, which uses a combustion engine to drive an electric generator-motor combination, also started with switching service and expanded to mainline freight and passenger service in the mid-1930s. Both new locomotives provided more and cleaner power than the steam engine, which died a gradual death in the United States after World War II.

It took about sixty-five years to go from a steam engine on a lab bench to a mature steam locomotive on the rails. In another sixty years, end-game improvement of the steam locomotive was underway, and its successor was entering the field. John Bockris, the originator of the photovoltaic-hydrogen concept, notes that sixty years is the typical time for a technology to move from invention, through initial engineering studies, to full commercialization. Successive energy technologies seem to have enjoyed another sixty to eighty years of unchallenged commercial dominance. This was certainly the case for water-powered and then for steam-driven machines. Electricity from fossil-fuel generation rose in the late nineteenth century, and a competitor, nuclear power, emerged in the 1960s. This sixty- to eighty-year lifetime does not represent immutable law, but it is important to understand that no technology represents the end of development.

When the gasoline car is examined in this historical context, its replacement seems timely. The car matured before 1930, less than seventy years after the internal combustion engine had been invented. It has had sixty more years of commercial success, largely unchallenged. So, in a historical sense, its gradual replacement over the next few decades should not seem premature. In addition, there are some indications that the internal-combustion engine is headed the same way as the steam engine, because an era of "compounding" is upon us, and once again the forces of change are the need for energy efficiency and pollution control.

The trend to more efficient engines has been underway for fifteen years; the drive for cleaner ones, for more than twenty. Turbochargers and superchargers in cars perform a function analogous to compounding in steam engines: boosting efficiency by extracting energy that would otherwise be exhausted. Adding air-pollution control equipment, like compounding for efficiency, extracts less benefit with each successive accretion. Eventually, a total redesign will become inevitable.

In the meantime, the days of gasoline (and diesel fuel) are numbered. American oil production is declining, and advanced extraction methods will be needed shortly to utilize what remains underground. The shrinkage of global reserves, to say nothing of instability in the Middle East, will inevitably drive prices upward well before the middle of the next century, accelerating the move to efficiency and to substitute engines, fuels, and transport modes. If global warming is to be taken seriously, the search for substitutes should not lead to coal-based synfuels, like methanol, for they produce twice as much carbon dioxide as oil. Again, the issue is how this natural process may be accelerated to take into account the exigencies presented by the greenhouse effect.

This line of thinking leads directly to three questions: What will power a future car, if not a gasoline internal-combustion engine? What can be done to cut emissions while the new technology is perfected, a process that could take as long as sixty years? And instead of expending efforts on alternatives to the Lenoir–Otto engine, should the individual car be replaced altogether?

New Car

Earlier, we used a constructive technique based on a few desirable properties to design a potential energy-delivery system for the future. Let's put a twenty-first century car together in the same manner. Ideally, it should have no carbon-dioxide or smog-inducing emissions, while retaining high performance, long range, and easy refueling. There is only one design now being considered that satisfies these requirements: the hydrogen car. If hydrogen were derived from electrolysis of water using electricity originating in photovoltaic arrays, neither engine nor fuel production would emit carbon dioxide, or carbon monoxide, or hydrocarbons.

121

The idea of using hydrogen to drive engines dates to 1820, but practical application began in the 1930s, when Rudolph Erren used hydrogen vented from dirigibles to run the airship engines; he also put 1,000 cars using either hydrogen or hydrogen-gasoline mixtures on the road in Europe. The 1960s and 1970s saw the development of several prototypes, including one by the Perris Smogless Automobile Association of California. But the vehicles never caught on because, among other reasons, hydrogen was expensive.

With a few adjustments, hydrogen can power today's internal-combustion engine. But if the hydrogen car were to become popular, an entirely new motor would have to be designed specially for it. When hydrogen is used in today's internal-combustion engine without any catalytic controls on the tailpipe, it generates nitrogen oxide at levels about the same as those achievable on today's gasoline engines using the best available controls; tuned for superlean burning, hydrogen engines are expected to produce much less. The performance of a hydrogen-fueled engine is about the same as that of a gasoline engine, and the cost of producing fuel could quickly become reasonable if the more optimistic expectations of progress on solar-cell prices turn out to be justified. For example, Princeton's Robert Williams and Joan Ogden calculate that by year 2000 a photovoltaic-hydrogen car would be cost-competitive with the gasoline engine if gasoline prices rose to about $2 a gallon, which is already less than it costs in Europe today.

The problem with hydrogen is in the tank. Hydrogen may be carried in three forms: liquid, gas, or combined with other metallic elements in a solid called a hydride. In each case, less energy is available from a given volume of fuel than from gasoline, and special tank materials are also required. As a result, to have the same range as a gasoline vehicle, a hydrogen vehicle would need to have a larger and heavier fuel tank. The cost of the tank is highest for hydrides, and its size would be largest for gas; but liquid hydrogen has a unique problem. It must be maintained at −423 degrees Fahrenheit. After a few days, even a superinsulated tank will warm, causing the hydrogen to boil. As a result, the excess gas must be vented safely, resulting in a costly loss of fuel. Furthermore, liquefaction is expensive in the first place; it would raise the break-even price with gasoline from $2 to more than $3. To travel the same distance as a

122

gasoline car, a liquid hydrogen car would need a tank six times as large. Even so, the fully fueled weights would be about the same because hydrogen is much lighter than gasoline.

But liquid hydrogen has performance characteristics closest to gasoline: it is particularly low in nitrogen-oxide emissions, and, like gasoline, it can refuel a vehicle in only a few minutes, quicker than the hydrogen gas or hydride options. The purchase price of a hydrogen car would be about the same as a comparable gasoline car. The tank size and fuel cost could be cut by developing high-mileage vehicles that would permit a range of 250 miles without a monstrous tank. This goal may be relatively easy to accomplish because a combustion engine specially designed for liquid hydrogen promises to burn much more efficiently than a gasoline engine, though no one has yet built one. (The tank volume noted above is based on the use of modified gasoline engines.) Should engine redesign succeed and should research solve the boil-off problem, liquid hydrogen would become the fuel of the future.

In case the hydrogen car seems like science fiction, consider the following: a consortium consisting of BMW, the electronic giant Siemens, a Bavarian utility company, the aerospace company Messerschmitt-Boelkow-Blohm, and the Linde gas company is building a photovoltaic-hydrogen plant with 50 percent German government financing. The hydrogen will be used commercially in mixtures with natural gas. Some of it might be used to fuel BMW's 745i prototype hydrogen car, which looks just like a regular BMW. Otherwise, the consortium might sell some to Daimler-Benz to fuel hydrogen test vehicles that company has developed, or sales could be made in specialized niche markets such as urban fleets or airlines. The Soviets, for example, recently flew a TU-155 commercial jet with one engine fueled by hydrogen. And as for safety in handling and distribution, West Germany already has a 130-mile hydrogen pipeline network around Düsseldorf to transport the gas to various industries.

Old Car

The major competitor to a fuel-efficient hydrogen car in the distant future will be the battery-powered electric car, a new version of the technology that died early in this century. Incredibly, today's few electric cars have the same power source as the ones eighty years ago, a lead-acid battery. With little commercial interest in the technology, no one has attempted to improve it. As a result, the vehicle has two critical shortcomings that a 1909 model also suffered: it has a range of only about seventy-five miles, and recharging it after a day's driving still takes all night. Furthermore, its low acceleration and limited top speed make its performance unacceptable for most uses. Nevertheless, several U.S. utilities currently use British-made vans with ranges of about fifty miles as maintenance vehicles.

In urban areas, the future looks bright for electric cars. There is a BMW small-car prototype that uses a new high-performance sodium-sulfur battery, and it has speed and acceleration characteristics similar to a gasoline-powered subcompact. The vehicle's eventual range is anticipated to be about 125 miles, and BMW expects to market it during the 1990s. More advanced batteries may extend the range. The optimistic estimate is that costs would be comparable to a gasoline car, even with photovoltaic charging of the battery. But recharge time remains a daunting obstacle for electric cars. This problem might be resolved technologically, with batteries that can tolerate a high-current zap without self-destructing, or by development of metal/air batteries that require no recharge, just easy hand replacement of the metal component. Alternatively, filling stations could provide battery-swap service. Until these changes occur, electric propulsion will be restricted to niches, albeit some rather large ones such as the urban second-car market.

Bridge Car

If the cars of the future are to be powered by hydrogen and electricity, most of the cars of the next two decades will still be powered by gasoline. Faced with that certainty, we again need a bridge to carry us into a greenhouse-friendly world; and once again, that bridge will be efficiency. American new-car fuel economy averaged 14 miles per gallon at the time of the 1973 oil boycott but rose steadily thereafter, to about 28 miles per gallon in 1987. The reasons were simple: gasoline prices shot up, and federal mileage standards took effect. But now progress has ceased as gasoline prices have tumbled, and federal requirements call for only 27.5 miles per gallon for 1990.

As of the mid-1980s, an economy-sized Ford Escort still used only 13 percent of its fuel energy to move the vehicle; most of its energy went out the radiator and the tailpipe. Off the road, however, auto companies are doing better. The actual mileage of on-the-road cars is considerably lower than the potential exhibited by experimental prototypes. General Motors, Volkswagen, Renault, Toyota and Volvo have prototypes that attain between 60 and 90 miles per gallon in the city and between 70 and 110 on the highway. Toyota's four- or five-passenger AXV gets 89 miles per gallon in the city and 110 on the highway. The adjustments needed to get such high efficiency include the use of lightweight body and engine materials, advanced transmissions, superchargers, computer controls, and new combustion designs.

The performance of these cars is similar to current road vehicles. For instance, Volvo's LCP 2000 goes from zero to sixty m.p.h. in eleven seconds, two seconds faster than the average 1986 U.S. car. The four- to five-passenger capacity of Toyota's AXV is adequate in size for the lion's share of the current market. Though generally lighter, efficient vehicles can be designed to be just as safe as standard cars because they incorporate new body designs and advanced materials that allow for both structural integrity and light weight. For example, Volvo's LCP 2000 withstands frontal impacts of 35 m.p.h., above the U.S. safety standard of 30 m.p.h.

Some of these high-mileage innovations, like weight and drag reduction, cut smog-generating emissions at the same time because they simply reduce fuel use. But other mileage-saving measures could actually increase conventional air pollution while reducing carbon-dioxide emissions because they also change combustion conditions: lean-burn engines, for instance, yield higher nitrogen oxide emissions.

The alternative carbon-based fuels—natural gas, methanol, and ethanol—have been touted as an air-pollution remedy, although the extent of the benefit from each is the subject of controversy. Among them, natural gas also yields less carbon dioxide than gasoline, but its low density limits driving range to about 100 miles. The carbon-dioxide benefit of renewable ethanol distilled from crops could be lost because of the energy expended in making it. Using methanol also limits driving range and produces no carbon-dioxide benefit at all. Furthermore, although it could be distilled from wood, methanol fuel in the United States is made from natural gas, and is generally imported. There remains considerable uncertainty over the extent of gas reserves; and should they become depleted by a shift away from gasoline to these alternative fuels, gas—and therefore methanol—would be made from coal in a synfuel process, and net carbon-dioxide emissions would skyrocket.

Instead of encouraging widespread use of alternative fuels of uncertain merit, efforts would be much better spent on incentives for design and production of safe, efficient, low-pollution gasoline cars as a bridge to hydrogen.* Done stupidly, redesign could yield top-mileage cars that are dirty and deadly. Done cleverly, it could produce cars that are much better than today's models in all three categories, while not quite doing the best in any one of them. And many of the efficiency gains invented for gasoline cars will pay off by giving us longer range, cheaper running hydrogen cars in the future.

No Car

The most intriguing alternative to new cars or efficient cars is a new paradigm altogether—no cars, or at least less driving. Before dismissing this option out of hand, recall that most of the world has not yet bought into the automobile society. The United States's ratio of more than one car for every two people dwarfs India's ratio of one car for every 500 people, or China's ratio of about one car for every 1,000. The United States has developed a monoculture in transportation, and the aforementioned socioeconomic framework has grown around it. The structure is extremely rigid: the car has no direct competitor so it cannot be excised.

The key to reducing dependence on automobiles is the gradual

---

*Depending on resolution of the air-pollutant emissions question, natural gas could reasonably play a limited role in fueling urban fleets. As we noted in the previous chapter, natural gas also can serve as a bridge to hydrogen in electric power generators like turbines and fuel cells, where it may be exploited with very high efficiency.

restoration of flexibility, and the key to flexibility is choice, or diversity. Fortunately for China, India, and other Third World countries, their societies are in rapid transition but have not yet congealed around the car.

The task of developing transportation facilities that minimize global warming will require a slow deconstruction of existing arrangements in the United States and avoidance of U.S.-type patterns of development elsewhere. In particular, the metropolitan model of core city and suburban sprawl is designed to maximize rather than minimize travel distances, and to minimize rather than maximize the effectiveness of mass transportation. One greenhouse-benign alternative would be based around a "polynucleated urban area," in the words of K. S. Kim and J. B. Schneider of M.I.T. Residential development would be clustered with employment centers, in contrast to the current situation in which zoning discourages such an arrangement. Several nuclei might radiate outward along corridors from a major urban center. Mass transportation, already more economical and energy efficient under some circumstances, could also become more convenient than using individual cars.*

From an optimistic point of view, this shift should occur in the natural course of events. Kenneth Jackson noted in 1985 that "the United States is not only the world's first suburban nation, but it will also be its last. By 2025 the energy-inefficient and automobile-dependent suburban system of the American republic must give way to patterns of human activity and living structures that are energy efficient." Jackson was writing when oil prices were still relatively high. From today's vantage point, if such a process is in the offing, it will happen all too slowly unless government moves it along intentionally.

If transportation is to be rationalized to avoid global warming, some lessons from the past must be heeded. Primarily, settlement must be planned in order to allow alternatives to the automobile to flourish instead of letting the automobile determine the pattern of settlement. This means supporting mass transit instead of gradually

---

*In the longer term, power for rail systems would come from hydrogen fuel cells, either centralized or carried on board.

encouraging its demise, as the United States has done ever since World War I. In developing countries, it might mean reserving rights of way now and laying tracks later, for example.

Until recently, no serious effort had been made to deconstruct an existing megalopolis gradually and reconstruct it in a different image. Now the Los Angeles region is flirting with modest moves in this direction in response to its air-pollution crises. The South Coast Air Quality Management District and the Southern California Association of Governments have drafted a unique plan aimed at cutting smog levels by about two-thirds in the face of expected growth of 37 percent in population and of 68 percent in projected vehicle use by 2010. Recognizing that the car is in Los Angeles to stay, the plan advocates a shift to electric or other nonpolluting vehicles after 2007. The plan's authors envision innovative approaches which would serve as a bridge to battery cars, like electrifying some freeway lanes to permit recharge during driving, and a private-public partnership to spur commercialization of electric vehicles. But the plan itself relies heavily on those problematic alternative fuels, methanol and natural gas, for the near future.

Nor does the plan address the fundamental question of whether growth itself should be discouraged or shifted elsewhere. But it does go far beyond the usual measures like bus lanes and car pools. The plan recommends land-use controls that would encourage a modest alteration of the present pattern, which forces long commutes in the car, by using zoning to decrease the separation of employment from residence. Some of the projected new job growth, 9.5 percent, would be directed toward housing-rich areas, while 4.2 percent of new housing would be sited in job-rich areas. These would be the first steps toward converting the area from sprawled to "polynucleated."

Attacking the structure of metropolitan areas really gets at the root of the entire transportation problem. Urban driving is disproportionately troublesome because its stop-and-go character leads to low fuel economy and as a result high emissions of carbon dioxide and other pollutants per mile. Yet if metropolitan areas were built to accommodate mass transit, and if small electric cars were also available for city driving, then people would have little need for their own gasoline cars. For intercity travel, they might consider using high-speed trains or rental cars, instead of continuing to own a high-performance car just for these infrequent trips.

The prewar dream of urban planners for "balanced" transportation, for diversity, is the key to building a benign system. There is a reasonable chance that this balance may be achieved in the Third World, where despite a budding romance with the automobile, there is still time to embed a different structure. But the United States is in trouble on cars. Even the modest land-use shifts envisaged in the Los Angeles plan are running into opposition from towns that want revenue from one use or another or just don't want mixed-use zoning. During the 1970s, several attempts by the federal government to encourage community land-use planning failed miserably due to a political backlash. Yet if we are unable to make this kind of adjustment, it will eventually hamper our economic competitiveness. For example, if high-speed intercity trains become a dominant mode of transportation elsewhere in the twenty-first century, the United States could be a nonentity in a burgeoning global export market for the technology. Imagine the United States without its flagship airliner industry, and you see the magnitude of the lost opportunity.

Many Americans love individual high-performance cars, and they can still have one, a hydrogen car, only with a higher fuel cost than today. By then gasoline will be more expensive, too. But this matter of taste need not continue to be the design basis for our entire society. Freedom follows from diversity, so the car monoculture is really a tyranny. Don't force people into cars. Give them diversity—some small electric vehicles, some hydrogen cars, some mass transit—and a flexible, greenhouse-benign world can slowly be put together over the course of four decades: just in the nick of time.

THE LONG ROAD

All technologies become obsolete eventually. The end of the gasoline car is coming, but a little shove is needed. What can government do? It should encourage the redesign of cities so that cars are needed less. Such an adjustment is easiest in areas that are growing;

but all places change, and change can be guided. It could start procuring electric vehicles for its fleets while seeding research on hydrogen's fueling problems and on quick-recharge batteries. It could stimulate efficiency with mileage standards and fuel use fees because efficient gasoline cars are a bridge, and efficient hydrogen cars will have a longer range. Meanwhile, the automobile industry must begin to think about air-pollution control as a redesign opportunity that can be accomplished simultaneously with efficiency, rather than as an unwelcome occasion for more compounding.

The need to act is desperate because the Third World and newly industrializing countries like South Korea are quickly getting addicted to the U.S. way. There are now nearly 400 million passenger cars in the world and an uncertain number of trucks; all modes of transportation together consume more than one-third of global energy. To see how pernicious the process is, note that merely to compensate for the increasing carbon dioxide emitted as Americans drive more and more would require raising fuel economy here to about forty miles per gallon by 2000 and more thereafter.

Should individual gasoline cars come to dominate the rest of the world as well, the greenhouse problem will become intractable. We need a new car and a new system. To speed the future, perhaps we should emulate the past, when the innovations of a thousand small shops catapulted us into the automobile era. Now innovations must lead us out of it.

CHAPTER 8

Eye of the Tiger:

The Alternative Path for Third World Development

THREE-QUARTERS OF THE GREENHOUSE GASES ARE GENERATED BY only 20 percent of the world's population, living in North America, Europe, and the Pacific Rim. Most of the human race, however, is found in Africa, Asia, and Latin America, the so-called Third World. Some of these nations, such as Bangladesh and Ethiopia, are contributing little to global warming. Already living on the edge, they will be the first to be overwhelmed by its consequences. But others, like China and India, are developing rapidly. If current trends remain constant over the next forty years, a steadily increasing proportion of emissions will come from these countries, as their populations expand and as their standard of living improves. Ultimately it is in the Third World that the fate of the Earth's climate will be decided.

PROBLEMS AND POSSIBILITIES

The Worst Case

Rising in the Himalayas, the Ganges and the Brahmaputra rivers

follow a north–south notch in the Earth's surface between the Rajma-
hal Hills and the Garo Hills northeast of Calcutta, eventually reaching
the Bay of Bengal atop the Indian Ocean. A hundred miles north of
the coast, they join the Meghna River, which drains the tribal hill-
states of India to the east. These three rivers, their thousands of channels, and the sixty feet of gray and red sedimentary soil built up over
eons of annual flooding, form the river delta known as Bangladesh.

With the 1947 partition of India, Bangladesh became the back-
water eastern wing of a geographically divided Pakistan until the
bloody civil war of 1971 brought it to the attention of the horrified
world. After gaining independence from Pakistan with the help of
India, Bangladesh slid back into obscurity, its presence as a nation
marked in the West by news stories about the cyclones that tear
across its southern coastline each year, sometimes drowning
hundreds of thousands of people at once. Today, Bangladesh
remains poor and vulnerable. It will be in the vanguard of victim
nations as Earth warms and the sea rises.

Much of the havoc waiting to be wreaked on the rest of the
world is already evident in Bangladesh. If global warming is slowly
nudging us over the cliff, Bangladesh has a head start into the abyss.

With a population of 112 million people packed into an area smaller
than Wisconsin, nearly every square inch of arable land is cultivated,
including the here-today-gone-tomorrow sandbars stippling the wide
rivers from the capital of Dacca down to the Bay of Bengal. This entire
flood plain was once a tropical forest. Now it is checkered with paddies
of rice and jute. As more land has come under the plow, forest cover
has decreased from 16 to 6 percent in just the past twenty years. The
last vestige of real wilderness is the Sundarban area in the southwest
corner of the country, where a forest of mangrove trees, home to the
last of the wild Royal Bengal tigers, fringes the coast. But neither the
mangroves nor the tigers are long for this world, because the entire
area lies within three feet of sea level. As the world warms, the Sundarbans will disappear; and this drama, a preview of nature's future,
is well under way.

Hard by the Sundarbans to the north is the village of Dhangmari,
a rickety collection of shacks woven from palm fronds and an occa-
sional wooden slat. The people, half of whom are Muslim and half
Hindu, scratch out a bare living by growing rice, fishing, and culti-

vating shrimp in flooded fields. They have a life expectancy about half that of Americans, and they die from the sorts of diseases, diarrhea for instance, caused by poor sanitation or a lack of shelter. The groundwater is too salty for consumption, so people collect their drinking water from surface ponds that fill during the rainy season and then fester for months afterward, breeding insects and disease.

Bad as the situation was, it worsened in 1974 when India closed the Farakka Dam on the Ganges. The southwest corner of Bangladesh dried up for half of the year. Formerly, a strong surge of freshwater flowed downriver toward the Sundarbans, blocking an incursion of saltwater moving upriver from the Bay of Bengal. After the Farakka Dam was built, saltwater moved upriver to foul the rice paddies that had been good for two crops a year. Now during the dry season the paddies are too brackish to cultivate; and as the sea rises in the future, saltwater will intrude farther inland, and more paddies will become sterile.

The Farakka Dam may be causing other problems in Bangladesh's two-front battle against flooding. In September 1988, water rushed down from the stark, partly deforested Himalayas of Nepal and India. Bangladeshis claim that intentional releases by India when the dam threatened to overflow aggravated the situation. The flood killed at least 1,000 people, made 20 million homeless, and did almost $1 billion in damage.

The ocean lapping at the Sundarbans promises bigger troubles than saltwater intrusion. Cyclones generated in the Bay of Bengal frequently make landfall at the Sundarbans and then swing eastward in a broad arc to batter the entire coast. In 1970, perhaps 200,000 people drowned in one of these storms, or maybe it was as many as a million. No one knows for sure, because many people were eking out a bare existence on transient sandbars where the rivers enter the bay, and there was no census of them before the storm, or following it either.

In November 1988, a cyclone struck the village of Dhangmari. Storm-driven waves couldn't make it through the protective mangrove forest, but the hundred-mile-an-hour winds did. Every frond-woven hut collapsed, and villagers escaped death only by swimming through chest-high waters to safety. Saltwater destroyed the remaining rice crop, government assistance was nil, and the only help came from private relief workers. The very simplicity of the houses allowed survivors to reconstruct them quickly, but as of early 1989 children had yet to return to the school because it had not been rebuilt.

Dhangmari has yet another problem: it sits three feet above mean high tide, but just as the sea is rising, the land is simultaneously sinking from tectonic movement. As a result, the Indian Ocean may race through the Sundarbans twice as fast as the sea will inundate the world's other coastal areas, permanently drowning Dhanghmari in perhaps forty years by some estimates, eighty years by others. The Farakka Dam only speeds the process because sediments carried by the river previously fed new land to the delta to compensate for the tectonic sinking. Sediments that should have raised the height of the Bangladeshi plain now lie instead behind the dam in India.

Dhanghmari has no future in the greenhouse world. The rising sea will take away the mangroves and salinize the paddies. The sea will eat into the coastline and tropical storms, amplified by warmer ocean water, will penetrate ever farther into a landscape afforded even less protection by the dying mangroves. New trees will not be planted because a burgeoning population, already at the edge of survival, won't cede land now used for cultivation, and so more people will die as each storm lashes the low-lying coast. If a Royal Bengal tiger manages to survive extinction, it will have nowhere to go but to a zoo.

The destruction will not stop at Dhangmari. Behind it lies the city of Khulna with more than 100,000 people. Behind Khulna lies the densely populated capital region surrounding Dacca. Much of the land between Dacca and the sea, a quarter of the country's territory, home to 30 million people, is sinking, and little of it stands more than nine feet above the Bay of Bengal.

Where will these environmental refugees go to escape the rising sea and the cyclones? What does an already overcrowded country with a current population density equal to Phoenix do when it loses a quarter of its land?

How can Bangladeshis even contemplate such a problem, crippled as they are by a litany of others? For instance,

- The daily wage of sixty cents is a fraction of the cost of a beer at Dacca's Sheraton.
- Most of the annual 4-percent economic growth occurs at the top,

134

and much of it is rumored to wind up in Swiss bank accounts rather than internal investment.

- More than one-third of the nation's budget is spent on the military.
- Eighty percent of the population can't read or write.
- Women still bear more than four children, on average.
- Mobility is stymied by the lack of roads on the wet terrain, and even Dacca crawls to a standstill with rickshaw traffic jams that outdo Los Angeles.
- A ten-year-old boy who can even manage some English has to quit school to sell peanuts at a rural bus stop to support his family.

There is a temptation to write off Bangladesh as a dead loss, to regard sea-level rise as the final blow to a very sick patient. As Rafiq Khandoker, a geologist at the University of Dacca, mused while contemplating his own prediction of a major earthquake in the country within a century, "Why worry about a big earthquake in 100 years if half the country could be under water in 40 years?"

But Bangladesh is also a land in which blind Gangetic dolphins leap in quick arcs from the rivers into soft, cool air and gentle light on a winter morning; a country of otters and spotted deer, of soaring eagles and gaudy parrots. It is a country of diligent, friendly people whose courage and flexibility in the face of adversity find no match at all in the West. Khandoker, for one, sheltered thirty-two relatives in his two-room house during the devastating floods of September 1988.

It is also a country in which, for all the poverty and desperation, some people worry about environmental pollution. During one week in February 1989, the *Dakha Courier* magazine discussed the results of garbage-incinerator toxicity tests carried out in Philadelphia, while newspapers reported on two upcoming conferences in Bangladesh on the impacts of sea-level rise from global warming.

A visitor sees some surprising indications of the potential for rapid change. In Khulna, the proliferation of video-cassette parlors stands out in remarkable contrast to the dearth of other amenities. These parlors could do much to disseminate useful information in a country with a staggering illiteracy rate. When A. Atiq Rahman, a development expert from Dacca, says, "Let's do the right thing on emissions even if the greenhouse theory proves to be wrong. Let Bangladesh do its small part," a listener believes that the country is prepared to try. And, in spite of its shrinking forests and meager reserves of coal (fossil energy comes from oil and natural gas, some of it local), Bangladesh may be able to help. Of worldwide methane emitted by cattle, 3 percent comes from Bangladesh. In addition, the country cultivates about 5 percent of the global rice crop, which also gives rise to methane. New cattle-feeding regimes and new rice cultivars might cut these emissions, and accelerating the recent downward trend in the birth rate would also contribute to the global effort. If the Bangladeshis are prepared to take such steps, surely there are no excuses for inaction by others.

In the future it may become possible to agro-engineer reductions in methane emissions. But it remains unlikely that Bangladesh or any other poor country can afford to initiate the requisite research. Such innovation, or at least the funding for studies, must come from abroad, particularly the industrial countries, which are themselves big methane emitters. American cattle are second in total farts only to Brazil's. Yet, as an old Bangladeshi saying goes, "If a man has enough money, he can get the eyes of a tiger."

The uncontrollable flooding of coastal areas, salinization of cropland and drinking water, brutal and destructive storms, people fleeing disaster zones, the inability of any one country to slow the rising tide: these are the images that will loom throughout the greenhouse world. Bangladesh is not just a poor relation we'd prefer to ignore. Bangladesh, in many ways, is the future for all of us. In a sense, global warming may make us all Bangladeshis.

Sun and Dung

India has particular cause for concern if Bangladesh's coast moves inland, for one possible answer to "Where will the people go?" is "India." To escape political turmoil, thousands of Bangladeshis have already crossed the Indian frontier into Assam State, even though some have met with violence and death. This trickle threatens to turn into a torrent of refugees over the next few decades if warming is permitted to proceed. But if the Earth warms and the sea rises, India will share the blame as well as the burden.

Two potential development paths coexist side by side in India. Both are intended to lead to the same goal: to lift per-capita income far above the present $290 a year (twice that of Bangladesh, but only about one-sixtieth of that of the United States), enabling the population to buy middle-class amenities such as television sets and refrigerators. To achieve it, India must generate far more electricity than the 55,000-megawatt capacity of its operating power plants (about twice as much as New York State). For example, an additional 80,000 megawatts would be needed if every family had a TV set, and television is only one of many reflections of the economic growth desired by Indians.*

One path to development assumes that the new electricity will largely come from fossil fuels, in which case such an increase in television ownership alone would boost India's current carbon-dioxide emissions by more than half. The other path becomes apparent to a visitor the moment he steps from the airplane: the Indian sun beats on the head like a hammer, and the nose recoils from a smoky-sweet haze rising from dung fires. Later, one notices the bio-gas digesters, photovoltaic modules, and solar cooking stoves that are springing up in the villages. The subcontinent averages 210 days of very bright, very direct sunshine each year, about six times that seen in Europe, and it is the sun and dung that are India's, and the world's, best hope for the future.

Mahatma Gandhi, "the Great Soul," would have loved solar energy. Indrashanbar Raval, who was with Gandhi at his ashram at Ahmadabad, is a research officer at the nearby Vishalla Environmental Centre, a small museum and energy-demonstration project. Raval points out that little coal (only 1 percent of the global reserves) lies under India, though it ranks high (fourth) in coal use, and he compares this situation to Gandhi's lesson about cotton. India exported cotton goods until the British took control of the raw material, sent it to mills back home, and forced India to import textiles. Arguing that economic independence was a necessary con-

*Although Indian cities are crammed with billboards advertising television sets, many villagers are still without electric lights to read by.

137

comitant to political freedom, Gandhi adopted the spinning wheel as his symbol and urged Indians to relearn the lost crafts of spinning and weaving.

Solar energy is like cotton to Third World countries: they have the raw material in abundance, and if the infrastructure for conversion is developed, they can achieve complete independence from imported energy. In fact, they may become profitable exporters of solar hydrogen to the darker north.

Aside from sunlight, India has another resource in abundance: cow dung, long used as an indirect solar-energy source. The sacred cows of the Hindus wander freely around the towns and fields. Brick walls in cities and countryside are plastered top to bottom with even rows of neatly molded discs of dung drying in the sun. Because dung derives ultimately from carbon-containing pasturage, which is renewed again and again, burning it provides energy without increasing carbon dioxide in the atmosphere. Traditionally, the cow pies are burned in inefficient open-fire dried-mud stoves. But now, with the help of the government's Department of Non-Conventional Energy, villagers are using a new method.

Bacteria feeding on fresh manure in tanks, called digesters, convert some of the manure's carbon into methane, which is then burned directly like natural gas. The advantages of this process are multiple: it captures more of the energy content of the dung while the solid sludge residue of digestion can be used as fertilizer; methane that would otherwise rise from moldering dung piles in the fields and contribute to global warming is put to good use, replacing nonrenewable energy; and the objectionable dung smoke is eliminated completely because the minerals that form smoke when dung is burned directly are now left behind in the sludge. Methane turns to carbon dioxide when burned, but the cycle is closed when more feed is grown for cattle, removing equivalent carbon dioxide from the air.

Saroz Bale is a housewife living in the small village of Masudpur on the outskirts of New Delhi. Luckily for her, an experimental nonconventional energy facility sits next door. Gas bubbling from dung simmering in large digester tanks is piped to her kitchen for two hours in the morning and two hours in the evening. She finds the small gas burner resembling one atop a standard range more convenient than its predecessor, an outdoor open fire in a baked-mud ring. The residual sludge is dried, packaged, and sold for fertilizer. Most of the cooking needs of 140 of the village's 180 houses and 1,200 residents are satisfied by the gas, the balance by dung fires, wood, or liquid propane gas.

Larger facilities like Masudpur's are convenient for institutional customers or whole communities, but biogas digesters in rural areas are usually scaled to household use. A family with three or four cows can satisfy its cooking needs with a digester out back, and a million of them are currently in operation.

The Department of Non-Conventional Energy has other efforts under way involving solar cooking stoves, photovoltaics, wind farms, solar ponds, passive solar construction, and solar thermal-power plants. About 100,000 aluminum-box solar stoves, simple sunlight reflectors, have been manufactured. Though slow to heat, the stove can boil two quarts of water at a time. Moreover, the department has equipped several dozen "energy-complex" solar villages with photovoltaic modules manufactured at the government's own amorphous silicon production facilities. All told, the renewable-energy program receives about $60 million a year in funding and is expanding, compared with the $112 million proposed in the much larger 1990 U.S. budget. As a fraction of gross national product, India's growing public investment in renewables is equivalent to ten times the United States's public investment, which has declined sharply since 1981.

Nevertheless, there are difficulties with India's energy program. The solar cookers cost about $25, even with a 50-percent government subsidy. That's about 10 percent of the average annual family income. There are ads for the stoves in the market in Old Delhi, but the extent of their use in the villages is uncertain. A cheaper but less efficient stove could be made on site with aluminum foil, but foil is not readily available in the villages. Similarly, a small biogas digester may be convenient for the four-cow family, but these families are in the top quarter of the income scale. If the shift to renewable energy continues to make the "wealthy" richer, it may become politically unpopular. Finally, when compared to other governmental efforts, renewable-energy expenditures are just a drop in the bucket. For comparison, World Bank loans running into the billions of dollars have supported construction of 8,000 megawatts of coal-burning power plants over the past decade. And in the next decade, the Indian government plans to spend about twenty times more money on nuclear energy than on renewables.

Still, programs are in place that attempt to put renewable energy directly into the hands of villagers who have limited access to other energy. Street and household lights, communal pumps for water, and communal refrigerators for medicines, all powered by electricity from renewable sources, are inching their way into rural life. Small industries may follow. Most important, the program's direction is toward expansion, and the philosophy is to exploit all options for future energy supply.

India's population may grow 20 percent by the year 2000, topping 1 billion. If the first development path is pursued, with fossil fuels relentlessly exploited, its carbon-dioxide emissions, already releasing 150 million tons of carbon per year, will simultaneously double. But poor countries have generally paid less attention to efficiency than even the United States because less capital is available for initial investment. In India the cooking process, which usually uses wood or dung, is only one-third as efficient as cooking in the United States and less than one-tenth as efficient as in West Germany. Electricity wasted in transmitting power from one place to another averages 22 percent, perhaps double U.S. losses. Professor J. M. Dave of Nehru University in New Delhi calculates that 40 percent of India's projected carbon-dioxide increase could be avoided by the dissemination of more efficient stoves, using biogas and a modest amount of direct solar energy, and by increasing the efficiency of industrial processes like steel fabrication.

Since the beginning of the Industrial Revolution, each nation's energy efficiency has increased as its economy matured. With the greenhouse effect bearing down on us, the trick is to accelerate this natural process by disseminating advanced technologies from the developed world. Obviously, halting the growth in carbon-dioxide emissions cannot be accomplished by such measures alone. It can only be slowed. But should the incipient photovoltaic industry catch fire, a fuller replacement of fossil fuels may emerge early in the next century, and emissions could be cut dramatically.

Full Throttle

While India's solar-energy program arouses modest optimism, pessimism quickly returns on consideration of China's breakneck exploitation of native coal. Propelled by energy use growing at 5 percent a year, China's gross national product is exploding at more than 7 percent annually. It doubled between 1980 and 1990 and will double again by year 2000. While China is rocketing toward industrialization, it is bloating global carbon-dioxide emissions because its population of 1.1 billion exercises high leverage on the worldwide emission total. Even a modest expansion of fossil-fuel consumption per person could overwhelm restrictive measures in the United States or Europe. And China's future energy plan is anything but modest. It calls for catching up to Western living standards by the middle of the twenty-first century (by providing every household with a refrigerator, for instance). And with more than a quarter of the world's coal reserves, China can thumb its nose at any attempts to limit fossil fuels.

As far back as the thirteenth century, Marco Polo noted that China used coal instead of wood—"all over the country of Cathay there is a kind of black stone existing in beds in the mountains which they dig out and burn like firewood"—but not until the 1870s did an explorer, Baron Ferdinand von Richthofen of Germany, become the first to reveal to the West the extent of China's great reserves of coal. The richest fields were in the north, but coal was found in every province in the country.

In the late 1980s, China surpassed the United States as the world's biggest coal burner and accounted for 10 percent of global carbon-dioxide emissions. By 2000, China hopes to burn almost twice as much coal, and by 2030, with a population of 1.5 billion, it may account for nearly 20 percent of global emissions, about equal to its share of global population. According to Professor Lu Yingzhong of the Institute for Techno-economics and Energy Systems Analysis in Beijing, only 10 percent of China's coal and oil is now consumed in the rural areas, where 80 percent of the population lives. Rural Chinese obtain 80 percent of their energy from fuel wood and agricultural waste, and this method slowly robs forests and soils of carbon and nutrients. But as the bulk of the population turns to coal, carbon-dioxide emissions will soar.

141

At this point, U.S. per-capita carbon-dioxide emissions are more than ten times China's, but if China were to achieve its economic goals by matching U.S. per-capita fossil-fuel use, its carbon-dioxide emissions would equal the total from the rest of the world today! If you think that China can't do this, consider that it is already ahead of Japan's 1914 energy use per capita. Even if energy use lags behind present U.S. consumption by more than half well into the next century, China alone would be responsible for committing the planet to another full degree of warming by 2030.

The alternative to this grim prospect is for China to base its economic growth on efficiency and renewable energy. Like India, China exploits part of its biogas potential by using pig manure in more than 4 million digesters for rural energy. Princeton's Robert Williams calculates that all of China's future electricity could be produced by burning gas distilled from wood in turbine generators at high efficiency, though it would take 14 percent of China's land to grow the fuel. The government is also developing small-scale hydropower dams, which may provide the electricity equivalent to twenty-five standard-sized American power plants by year 2000.

China, like India, remains fertile ground for increasing the gross national product through gains in energy efficiency because its economy is only about half as efficient as the United States's and one-fourth as efficient as Japan's. The Chinese government's plan incorporates a 2-percent annual efficiency improvement. As new capital stock is added, China could do even better without sacrifice of economic growth. For example, one-third of its steel production still comes from the inefficient open-pit hearth furnace, long abandoned by Japan, where steel production requires only half as much energy. One obstacle to such a change is the policy, common in the Third World, of keeping the price of energy below market rates. Just as artificially low food prices discourage farming, in parts of the Third World low energy prices discourage the "growing" of new energy through efficiency improvements.

The main thrust of China's energy policy is troubling; still, thinking could change. The planned rapid exploitation of fossil fuels may be difficult because, among other reasons, it remains uncertain that rail lines to ship the coal can be expanded fast enough. In contrast, efficiency in combination with renewable sources can help shift the burden to *local* energy. Without a large investment in pollution controls, moreover, coal exploitation will further degrade China's already poor air quality. Even now, Chinese are constantly choking, coughing, and spitting as a result of sulfur dioxide and particulate levels that are up to 100 times higher than those in the United States. Such considerations could make China ripe for renewables as a substitute for coal, but only if technological advance can lower their cost soon. Otherwise, the Chinese will continue to exploit coal to their own detriment and that of the rest of the world.

The Price

India and China are running toward the pot of gold at the Western end of the rainbow, yet they too have ample reason for concern over global warming. India is already too hot: summer temperatures in New Delhi can reach 115 degrees Fahrenheit, and climate models indicate that doubling carbon dioxide could push Delhi's summer temperatures an average of five to six degrees higher. (Of course, the heat is also one reason that Indians are eager to possess more air conditioners.)

Yet an hour south of New Delhi sit the Department of Non-Conventional Energy's passive solar buildings, in which the temperature remains below eighty-six degrees even on the most brutally hot days. It's not heaven, but it's reasonably comfortable. In March, with the outside temperature in the mid-eighties, under conditions in which a new office building in the West would be burning up electricity to stay comfortable, the buildings are cool. No air conditioning or daytime lighting is needed, just a few fans coupled with intelligent design and appropriate orientation of air vents and windows. Hot water comes from passive solar devices, which become an elegant part of the buildings' facades. If new construction in India incorporates these features, then future office buildings will rise without a commitment to heavy energy use. But if global warming arrives as forecast, the job of keeping buildings cool without air conditioning will become more difficult.

The problems brought to Asia by climate change will go well beyond mere discomfort. The rice crop, which provides the staple for much of the continent, is sensitive to temperature. Studies at the

143

International Rice Research Institute in Manila indicate that yields decline as heat increases and total failure can occur with prolonged exposure to temperatures above 106 degrees Fahrenheit. In addition, agriculture in southern Asia as well as much of China is sustained by the summer monsoon rains, which are produced by the seasonal shift in air circulation. Bulging into tropical seas, the East Asian land mass warms rapidly as summer approaches. As the hot air rises, cooler moist ocean air is drawn in over the land and pushed up over the Himalayan plateau. The moisture condenses at the high elevations and rains down over the rice paddies of China, India, and Southeast Asia. Failure of the monsoon can seriously affect rice production, as evidenced by the drought of 1971–72, when yields in parts of India tumbled. By the same token, intensification of the monsoon can lead to disastrous flooding of the sort that afflicted Bangladesh in September of 1988, which also caused disruption of the food supply. Results from at least one climate computer model suggest that the monsoon may intensify with global warming. But reliable projection of continental-scale circulation patterns is beyond the grasp of modelers, so the direction of change remains speculative. In any case, China feeds more than one-fifth of the human race on only one-fifteenth of the world's arable land. Under such marginal conditions, any change at all could be disastrous.

Nevertheless, some scientists with an interest in agronomy such as Paul E. Waggoner, along with Norman Rosenberg of Resources for the Future, a Washington, D.C. think tank, are optimistic about the future. They contend that the global food supply will be maintained except under the most rapid, dislocating warming conditions foreseen by the climate modelers. Farming will adjust or move to other places. But some Third World countries, like those in the Sahel in Africa, have difficulty supplying food to people even under today's conditions. When the weather is irregular, people starve. For instance, the poor harvests in the United States and the Soviet Union in 1988 drew world grain reserves down to 16 percent of their peak by early 1989, the lowest value in fifteen years. As a consequence, grain prices rose 40 to 50 percent, signaling better returns for exporting countries like the United States and correspondingly higher costs to importing nations. The UN Food and Agricultural Organization declared that fifteen countries, including Bangladesh, required assistance to avert starvation. A greenhouse future is bound to bring these countries more of the same.

The story of the delta in Bangladesh will be repeated, in a smaller way, at the Rann of Kutch on the west coast of India, at the Indus delta in Pakistan, the Yangtze and Yellow deltas in China, the Mekong in Vietnam and the Irrawaddy in Burma. The vulnerability to sea-level rise extends across the Third World to the Nile delta in Egypt, the Limpopo–Zambesi in Mozambique, the Niger in Nigeria, the Magdalena in Colombia, the Orinoco in Venezuela, the Amazon and São Francisco deltas in Brazil. Agricultural land, natural ecosystems, and inhabitants will all succumb to the rising seas. Populations will flee and political systems falter. If countries like China and India see the inevitability of economic expansion, they must also see the necessity of accomplishing it without fossil fuels.

DEBTS

Tropical rain forests are so thick with life of every kind that the intruder can almost taste the carbon. The air, moist and heavy, smells like the inside of a mushroom. Brazil's Amazon region contains nearly one-third of the world's remaining stock of rain forests, so important as a storehouse for carbon. Brazil also has a birth rate like that of India, one of the most skewed income and land-ownership distributions in the world, a life expectancy lower than China's, the biggest foreign debt in the Third World ($110 billion), and, not surprisingly, the highest rate of deforestation anywhere. With a population of 154 million people, the world's ninth largest economy, a wealthy elite, and abundant natural resources, Brazil is also a member of the emission-growth group with India and China. But for now at least, Brazil's greenhouse contribution comes largely from the subsidized destruction of its forests in an ill-conceived attempt to add to its agricultural lands. Other countries engaged in extensive deforestation include Mexico, Zaire, Nigeria, Indonesia, Thailand, and Malaysia. Some of these nations, particularly those in Southeast Asia, serve as Japan's private tree farms and deforest for timber rather than for land.

In the process of deforestation for land clearance, old trees are burned, releasing stored carbon dioxide to the atmosphere. Lumbering is, on the whole, less problematic than agricultural burning, for the carbon can sit harmlessly in a building's walls for decades. But clear-cut operations so disrupt the forest floor that the soil itself decays, releasing methane and carbon dioxide.

In 1988, worldwide deforestation pumped more than a billion tons of carbon into the atmosphere, approximately one-fifth of the amount emitted from fossil fuels. Brazil alone generated nearly as much carbon dioxide from forest fires as China did by burning coal. Conflagrations in the Amazon spewed out not only carbon dioxide, but methane, carbon monoxide, and nitrous oxide. The massive smoke plumes from these fires sparked another sort of firestorm, an international protest. More significantly, deforestation is now becoming an internal issue in Brazilian politics. These pressures led President José Sarney to propose a nature-protection program and to suspend the most flagrant incentives that subsidized the destruction.

Along with the tax incentives, which date from the 1960s, the policy for Amazonian development provided for road construction and colonization, which was abetted by the World Bank and other international lending institutions. Several motives lay behind these policies. The government wanted to increase land available for export crops and cattle to improve its balance of payments, secure sovereignty in border areas, and decrease pressures for the redistribution of large existing landholdings by continuously creating new acreage for cultivation by the landless. Access to timber and gold provided another lure.

This policy has been a financial disaster for the Brazilian government and an environmental disaster for the planet. Huge tracts go up in smoke as pioneers burn their way through the jungle. In just ten years, legalized arson, promoted by U.S. taxpayer contributions to international lending institutions, has destroyed more than 20 percent of the forests in Rondônia, a state that is as big as Oregon. Colonists soon discover that tropical soils are nutrient deficient and become rapidly unproductive. Typically, after two or three years, they sell out to ranchers and then slash and burn their way to a new piece of land, and more forest goes up in smoke. The ranchers fare little better. In the first year, every ten acres can support four head of cattle, but by the fifth year they can support only one. Meanwhile the cattle let loose to graze proceed to boost global methane levels.

But the death of the tropical forests does other harm besides contributing to the greenhouse effect. It destroys a genetic motherlode of millions of plant and animal species, most of them unknown to science. The product of millions of years of evolutionary development, these species have a right to exist, and from a human's "practical" point of view, their survival may be essential on medical grounds alone. Rain-forest plants comprise 70 percent of the 3,000 plants that the U.S. National Cancer Institute has so far identified as containing anticancer properties. Even so, less than 1 percent of all tropical-forest plants and animals have been screened for their potential uses of any kind. But for all their richness and incredible diversity of species, tropical forests are going up in smoke in an ecological holocaust without parallel on Earth. As William Beebe wrote, "The beauty and genius of a work of art may be reconceived, though its first material expression be destroyed; a vanished harmony may yet again inspire the composer; but when the last individual of a race of living beings breathes no more, another heaven and another earth must pass before such a one can be again."

Unless the human population growth is slowed, it will become progressively more difficult to protect forests and other resources in these countries. Even without destructive government policies, forests will be exploited in some places because, like China's coal, they are there. In Costa Rica—where land and income distributions are somewhat more even than in Brazil and an enlightened government is aggressively pursuing tourism with a farsighted program of nature reserves—forests still disappear at the rate of 4 percent a year. The resources do not exist to protect lands set aside as parks. Until human populations are stabilized and until the uneven distribution of land ownership and income is rectified in a country like Brazil, forests will continue to be leveled. Even a large army couldn't stop deforestation, though who knows what might be attempted by the northern countries if the greenhouse effect accelerates.

External pressures are arrayed against the forests, too. Several "deforesting" nations are so deeply in hock that one-third to one-half of their annual export income is applied to the interest on their debt. Far more money flows out of Latin America in this way than returns in new loans, and therein lies the pressure to turn forest land to growing cash crops. But where soils are poor, as in the Amazon, paying back the banks this way means irreversibly selling off major assets, natural resources, instead of managing them for a long-term income.

Until recently, the international lending institutions have hindered rather than helped in solving the deforestation problem. The World Bank initially supported Brazil's national energy-investment plan, which calls for construction of more than 130 dams by 2010, 79 in the Amazon. The bank's own analysis showed that building even half of the dams would cost $44 billion, yet investing $10 billion in improvements such as efficient electric motors in factories would make this construction unnecessary. No matter. Forests were to drown behind massive dams, while the capital raised to pay for the dams would bloat the foreign debt.*

Such thoughtless disregard for the forests, and the people who live in them, led the U.S. Congress to pressure the World Bank for change. In response, the bank has delayed some egregiously destructive loans and is finally paying attention to an alternative economic path called "sustainable development" advanced by the UN's World Commission on Environment and Development, known as the Brundtland Commission. Sustainable development is analogous to living off the income from your financial resources instead of eating up the principal (as deforestation and cattle ranching does to forests and soils).

The World Bank has joined with the Inter-American Development Bank in supporting this novel approach in the state of Acre next to ravaged Rondônia. With backing from the bank, pressure from grassroots organizations, and advice from the Environmental Defense Fund, Acre has established twelve "extractive reserves" in an area the size of New Jersey, where Indians and resident rubber-tappers will manage the forest in a sustainable manner by gathering fruit, oil, nuts, and rubber, some for export, without setting the trees ablaze. The local populations will reap a stable income exceed-

*During 1989, after considerable prodding by its American and European directors, the bank suspended support for Brazil's energy plan.

148

ing what could be gained from farming, and further greenhouse emissions will be avoided. The banks' support gives the government reason to protect local people's interests, and nondestructive activity has an opportunity to gain a foothold in the forest.

Then in 1988, Francisco (Chico) Mendes, leader of the rubber tappers' grassroots organization, was murdered, allegedly by the henchmen of landowners whose unrestricted freedom to expropriate land arbitrarily was threatened by the project. The bullets blew away the cover hiding the underlying social causes of deforestation and revealed an ugly reality. Protection of forests will be difficult because it is so tied to resolution of issues of economics and class. But global warming might finally give the lending countries and their banks reason to be concerned with the social problems underlying defo-restation, which they have long tried to ignore.

Debt-for-nature swaps, a concept originated by the Smithsonian Institution's Thomas Lovejoy, use the forgiving of debt as a tool to preserve lands instead of destroying them. So far only a few minor swaps, privately financed, have occurred; and unless the lender governments and major banks participate, these measures will not even dent the problem. Brazil, for one, resents the idea as an intrusion on sovereignty. One approach that averts this issue would be based on a pay-as-you-go system of forgiving a fraction of a year's interest for every year that an acre is preserved. Still, unless some money and opportunity trickles down to the average person, forests will remain in jeopardy no matter how clever the scheme. If people living in a forest are given rights to its renewable resources, they will have ample motivation to protect it. If landless people who are clamoring to get a little of the forest are given alternative opportunities, pressures will ease.

CONFLICTS

The Third World must have been taken aback by the new concern for its forests on the part of the United States after years of being encouraged to destroy them by U.S.-dominated banks. Perhaps it was this sudden change that in 1989 led Brazil's President Sarney to label U.S. pronouncements about deforestation as "an insidious, cruel and untruthful campaign" to distract attention from its own environmental degradation. Sarney had a good point, if we consider that northern nations do not always manage their own forests well, forests that are also important stores of carbon. Brazil is actually second in the world in total forested area: the Soviet Union is first, with Canada and the United States third and fourth.

Although the resulting contribution to global warming may be small, old-growth forests are still being exploited in the United States and the Soviet Union. One of the most egregious examples is the U.S. Forest Service's selling of the Tongass National Forest in southeastern Alaska. One of the world's last temperate rain forests, the Tongass is about the size of West Virginia. Under fifty-year below-cost contracts, the Forest Service is peddling 500-year-old Sitka spruce trees for about two dollars apiece to a pulp mill owned by the Japanese. The mill admits to earning a handsome profit shipping pulp to the Far East, where it is made into fabric for synthetic clothing.

Putting aside the environmental issues, the deal is an economic disaster for U.S. taxpayers. On the open market the trees would command around $300 each, but Alaska's two senators and its lone House member contend that small towns in the sparsely populated region would suffer if the government did not keep them propped up with the contracts. As long as this sort of cutting continues, exhortations from the United States, which decimated most of its frontier in the last century, will be certain to have a hypocritical ring to Brazil and other countries now exploiting theirs.

In the future, instead of encouraging deforestation, foreign assistance should be tied to environmentally benign development and land reform, much as currency loans from the International Monetary Fund are conditional on economic reorganization. Yet even if the Third World countries were to make planning for the greenhouse effect a priority, they are unlikely to achieve a meaningful restructuring without massive technology transfers from the West. As China's Environment Commissioner, Liu Ming Pu, said at the March 1989 Ozone Layer Ministerial Conference in London, "The developed countries consume 80 percent of the world's resources and produce an equivalent amount of pollution. . . . [Those countries] can use their past accumulated wealth to manage the environment. The Third World countries cannot do this."

India refused to join the Montreal Protocol initially and only grudgingly came to support (informally) a CFC ban in 1989 because it has its own CFC capacity, which it was reluctant to sacrifice. The substitutes would be more expensive to produce, and Western companies, which have had a head start in their development, would control access to them until their patents expired. Such inequities could destroy any hope of an agreement on greenhouse gases. If the West wants the Third World to use new technologies to replace fossil fuels, it will have to give the technologies away free or at least at bargain prices. Of course, the picture is complicated: the way things are going, the United States could be buying photovoltaics from India in fifteen years.

The same applies to the non–carbon-dioxide aspect of greenhouse gas emissions. New rice strains and cultivation methods and new cattle feeding regimes may be developed to cut methane emissions, just as alternative fertilizing schemes may reduce nitrous oxide. With its huge agricultural research establishment, the United States needs to encourage research on the solution to these problems and then be prepared to donate the findings to Brazil and other the big emitters.* Instead of employing agronomists to find heat- and salt-resistant strains of plants after the fact, let the agronomists create low-methane cultivars to avoid the warming in the first place.

THE OTHER ENGINE

Two engines drive up the emissions of greenhouse gases: one fueled by the desire for more food, goods, and services for each person, the other fueled by the increasing number of people who desire these benefits. The world's population was 500 million in 1500. It has expanded more or less continually to nearly 2 billion by 1900 and 5 billion today.

---

*Lower beef consumption in wealthy nations would help too, by cutting demand for cattle.

The historical pattern is changing now, albeit too slowly for comfort. The global population's growth rate increased until the 1960s but has declined ever since. This trend is most evident in the economically developed countries. Family size in Japan is small enough to have slowed population growth to a crawl, while the number of West Germans is actually declining. Projections envision the U.S. population peaking at roughly 300 million in 2030.

In a number of Third World countries, particularly China, India, and Indonesia, the population growth rate has also slowed significantly as the result of reductions in family size. But fertility rates remain substantially above the so-called replacement value of about two children per family, so population will continue to rise for the foreseeable future. In most of Latin America, population growth remains rapid, while it is exploding in much of Africa, where the average of six plus children per family hasn't changed much in twenty years.

The net consequence of these trends is a projected stabilization of the human population at perhaps 10 billion late in the next century. Some experts have optimistically argued that stabilization could occur a bit earlier at 8 billion people, while pessimists peg the figure as high as 14 billion, occurring some time in the early twenty-second century. No matter which of these figures is right, the rise in the number of humans will stimulate more greenhouse-gas emissions.

If the amount of fossil fuel needed to supply a given level of goods to each person could be reduced by efficiency measures, per-capita carbon-dioxide emissions would drop. But total global emissions might not drop as fast—they might even increase—as the population continues to grow. Superficially, it would follow that efforts to accelerate the stabilization of population would reduce greenhouse-gas emissions proportionately, but this view oversimplifies the situation.

Slowing down the population-growth engine might actually lead to a rise in each person's energy consumption, for it would result in more disposable income per capita that might well be spent on automobiles, electrical appliances, and the like. In fact, this is exactly what is now happening in China, where emissions increased by 57 percent between 1975 and 1985, while the population grew by only

14 percent. In a rapidly growing economy like China's, population growth is not the major force behind expansion of fossil-fuel combustion. The Third World's interest in limiting population—and almost all Third World countries urge family planning—is to provide more goods to each person, not to restrict energy use and economic growth. (Even in the United States, the thirst for energy remains unsatisfied. The U.S. population has been growing at less than 1 percent annually, but in the late 1980s energy use has surged at several times that rate.)

Nevertheless, it is inevitable that stabilizing population will help to slow emissions. Once the decision is made that it is imperative to control the emissions of greenhouse gases, implementation would be greatly facilitated by efforts to control population growth as well. No matter what happens, any move away from fossil fuels is bound to occur gradually. Whatever level of these fuels the world decides it can safely use over the intervening period will create more benefit per capita as long as fewer people need to share them. These people, in turn, will feel less of a pinch by the energy limits, which will make the restrictions more politically viable.

Moreover, population growth is directly responsible for the expanding methane emissions from cattle herds and rice paddies and pressures on the land that result in deforestation, soil erosion, and loss of wildlife habitat. Some authorities argue that even these abuses could be avoided without limiting population by more efficient use of resources and biotechnological advances in agriculture. If we are clever enough, the planet might well be able to accommodate severalfold more people. But in this world, on this limited planet, more people still means fewer natural resources. At the moment, it is only prudent to strive toward stabilization of population as soon as possible.

Arguments will seesaw repeatedly as the merits of various green house strategies are debated over the coming years. Third World countries with growing populations will castigate Western nations for using ten to twenty times more fossil-fuel energy per capita, while industrialized nations with stable populations will insist that the Third World's growth rate means fewer forests, as well as an unceasing pressure for more energy use, regardless of the environmental cost. Both sides are right. Both engines driving up the greenhouse gases need to be slowed. We need to distribute both compact fluorescent lightbulbs and condoms, though they alone won't solve the problem. Unless a higher level of economic opportunity and a concomitant level of literacy are attained, particularly among women, and unless adequate health care is available, even strong population-control measures will have limited effectiveness. For example, India's fertility rate appears to be stalled far above the replacement value despite stringent family-planning efforts in the past.

But even if everyone in the Third World attained a Western lifestyle based on efficient energy use and only moderate population growth, the planet would warm unacceptably. The only way to break this self-defeating circle is to develop renewable energy. Fast.

CHAPTER 9

The Fifth Wave:

The New Environmental Economy

THE SPECTER OF ENVIRONMENTAL CATASTROPHE LOOMS AT THE very height of widespread enthusiasm for laissez-faire governance. The global commons is screaming for protection precisely when the political and economic current has surged in the opposite direction in the United States, the United Kingdom, and parts of the Communist world. This approach has much to offer, as exemplified by the radical efforts of China and the Soviet Union to rejuvenate their economies; but all too often national leaders have forgotten that their role is to lead. The bungling, begrudging response of the Bush administration to the *Exxon Valdez* oil spill was a typical example of shrug-of-the-shoulders government.

In its extreme form, the free-market view amounts to this: a lasting solution to global warming will occur only when systemwide economic changes disfavor energy-intensive industries. These shifts will occur whether or not governments intervene and such intervention has often proved counterproductive. Therefore, why should governments intervene?

There is evidence to support this view. After all, the cost of photovoltaic power has dropped fivefold over the past decade, during a time of declining government support for solar energy, at least in the United States. It is conceivable that any remaining obstacles to

155

mass-market penetration will be overcome within the next ten years with no additional government assistance. The cost of photovoltaic electricity could gradually drop to the price of coal-based power, under optimal conditions, so that substantial substitution would begin.

Similarly, some nations are progressively using less energy to produce the same level of economic well-being. This trend has been slowed in the United States by lower energy prices, but it continues elsewhere, such as in China, and the trend toward less energy use can be expected to become universal again when the inevitable energy-price turnaround occurs, as happened after the OPEC oil squeeze in the 1970s.

Furthermore, the U.S. government has had a rather disgraceful record in energy planning. Congress spent $1.5 billion on the never completed Clinch River breeder reactor before killing the project in 1984. James Schlesinger, Jimmy Carter's energy secretary, pushed for a synthetic fuels program that would have disemboweled parts of the Rockies for shale oil and added to air pollution locally and carbon dioxide globally. Ronald Reagan's energy secretary, James B. Edwards, was no better. A booster of strip-mining, he insisted that youngsters in Kentucky would benefit because all the mountains leveled for coal could serve as ball fields.

As a resource drain, however, nothing matches the decades-long postwar effort to abet the light-water nuclear reactor industry. Great gobs of public and private money have disappeared into a black hole, while political energy and technical talent have been diverted from potentially fruitful pursuits, like accelerating renewable-energy development or designing cleaner and more efficient coal-fired boilers.

At the same time, we have shown in chapter 6 that the task of replacing fossil fuels and slowing emissions of other greenhouse gases is greater than simply developing photovoltaics or plugging in efficient lightbulbs. Although solar energy may sneak piecemeal into existing distribution arrangements, the unabetted transition to a mature photovoltaic-hydrogen economy remains farther in the future than the three-degree cliff. Likewise, the efficiency bridge is being assembled rather slowly in this country; in fact, its construction has come to a complete halt since 1985 when energy use surged again with lower oil prices.

Although it is true that computers, robotics, fiber optics, and composite materials will permit far more efficient utilization of energy in the future, and that innovations in the semiconductor industry will lead to improved solar cells, none of these changes is occurring fast enough to stave off global warming. Yet the arguments of the free marketeers should not be disregarded just because their conclusions are overdrawn. They offer two extremely important messages.

First, large-scale shifts would be occurring in the global economy that would alter energy-use patterns (as well as life-styles, work arrangements, and industrial organization) *even if the global environment were of concern to no one.* Second, they warn that governmental involvement in the process of guiding these changes is a two-edged sword. Government may inadvertently retard progress to stem global warming with wrongheaded choices, as witnessed by synfuels, the breeder reactor, and the light-water reactor. So what should government do?

THE MOTHER OF INVENTION

The global-warming solution as outlined in chapter 6 has two parts: an ultimate "new world," which is very different from today's in terms of energy, and a bridge to that world constructed largely of energy-efficiency measures, which provides a slow-warming path to get there. But an engineer surveying the gap to be crossed would conclude that neither the bridge nor the world on the other side can be built entirely from materials that are currently available and affordable. This observation is not comforting to anyone intending to cross over.

At the present time, the world uses about 80 trillion kilowatt-hours of fossil-fuel energy each year. This amount must be cut to about 60 trillion kilowatt-hours by the year 2000, and to less than 40 trillion kilowatt-hours by the year 2030 in order to stave off a three-degree rise. In effect, the total quantity of carbon dioxide emitted into the atmosphere must be reduced by more than half compared to today's releases. But the situation is even grimmer. World population will continue to grow, as will the level of industrialization of many countries. These factors will press emissions upward. To offset this trend, we need to reduce total emissions not just by one-half but by about three-fourths from where they are currently headed. If we do not, the chasm will widen just as the bridge is being built.*

Experts at the June 1988 conference, "The Changing Atmosphere," sponsored by the Canadian government's environmental ministry, concluded that, by implementing existing technologies, the world could just about meet the target for the year 2000 with no significant penalty to the global economy. But we need a considerably greater reduction to stop the warming. Where's the rest going to come from?

Some sort of contingency plan is needed. If an engineer needed a real bridge in place by a certain date, and if he were doubtful that the girders already on hand would be sufficient, he would order more girders for delivery when needed. Similarly, we need to order some new technology for the future in case the technologies on hand now won't quite span the gap.

HOW BUSINESSMEN THINK

How is it that in 1985 the world could not live without CFCs, yet by 1989 nations were preparing to abandon them completely? If one can understand this transformation, one can grasp the solution to the greenhouse problem.

Negotiations on the Montreal Protocol succeeded largely because of concerns about the environment: thirteen years of increasing confidence in the ozone-depletion theory and its implications for skin-cancer incidence, capped by the discovery of the ozone hole. But

---

*In order to avoid a three-degree warming, levels of the other greenhouse gases must be stabilized over the same period. Methods for obtaining reductions are discussed in chapter 10. Methane and nitrous-oxide levels may stabilize as warming slows, as noted in chapter 2, but large cuts must be made in CFC emissions.

there was another reason that was essentially economic: the availability of affordable substitutes (some immediate and others within the foreseeable future) and the potential profit to be reaped from them by a few big companies already in the CFC production market.

Cheap alternatives to the use of CFCs in aerosol spray cans were available more than a decade ago, but new refrigerants were slow in coming because public concern about CFCs eased after 1978. At that time, there was no measured ozone depletion, and, in any event, CFC emissions were decreasing following the aerosol bans. Industry does little or nothing for free, so Du Pont dropped its research and development on substitutes for CFC refrigerants in 1980. After all, who would want to buy higher-priced substitutes unless the government dictated their use, and why would the government do so if the public didn't care? As ICI Americas, a subsidiary of a British chemical giant, put it, "Further studies [of the substitutes] would require major investment in new plants to produce sufficient material," investments the companies were not then prepared to make. Bear in mind that industry was not just a passive observer of developments: the Chemical Manufacturers Association supported research on ozone depletion and exploited the uncertainties turned up by scientists to forestall regulation.

There is a chicken-and-egg problem here: industry won't pay for research and development unless it has a reasonable expectation that the fruits of research will find a market through the device of regulation, but the public won't demand regulation for a remote problem like ozone depletion unless there is a reasonable expectation of a cheap substitute or a looming disaster. In the case of CFCs, several events helped to break this vicious circle. The 1985 Vienna Convention, while not establishing new limitations, held out that possibility; by 1986, the ozone hole had heightened public concern; and then, with protocol negotiations imminent, Du Pont let it be known that, if given the right incentives, it could indeed produce ozone-friendly refrigerants at a reasonable cost within five years.

Du Pont's substitute was called FC-134a, a chlorine-free refrigerant that could replace CFC-12. Its timely appearance was due to Du Pont's earlier research and development program, which had originated in the 1970s when CFCs were identified as potential ozonedepleters. Had regulatory pressures remained after the U.S. aerosol ban in 1978, ozone-friendly substitutes might have become available even before the announcement of the ozone hole in 1985. From then on, scientific findings, public alarm, announcements of new substitutes, and finally regulation began to leapfrog each other. In September 1987, with a likely substitute in the offing, the specter of the ozone hole looming, and the Greens stirring up concern in West Germany, Montreal negotiators raced forward with an agreement. Shortly thereafter, James Anderson's results became public, and the Montreal decision to cut CFCs in half was already out of date.

In January 1988, AT&T and Petroferm, Inc. announced that a substitute for some of the solvent uses of CFC-113 had been developed, although, unlike Du Pont, they had kept mum about it until after the Montreal agreement was signed. On March 15, 1988, the NASA Ozone Trends Panel blamed CFCs for a 3-percent decline observed in global ozone levels. On March 24, Du Pont stated that it anticipated totally phasing out production of CFC-11 and CFC-12 within the course of the decade, because of the expected success of their research and development program. In January 1989, the company trumpeted a direct replacement for CFC-12 in air conditioners for cars already on the road. Then came the telltale signs of an ozone hole in the Arctic, which prompted a change in the European community's position that led to the May agreement in Helsinki to eliminate CFCs altogether.

The CFC-substitution question is a particular case of an old quandary: the solution to some problems depends on research initiatives in the private sector; but in a free market, companies are often reluctant to proceed with this research because sufficient incentives are lacking. We shall return to this point later. For the moment, it suffices to note that the resolution of this dilemma, discussed by economist Kenneth Arrow in a landmark 1962 paper, turns on governmental intervention.

Let us turn to the global-warming problem. The odds against strong regulatory action in the near future are high, and there has been no discovery as alarming as the ozone hole. There are some promising substitutes for fossil fuels, but they require more research and development. Consider some of the technical problems that must be solved to reach a "greenhouse-friendly" world. Amorphous silicon cells, now 12-percent efficient in the laboratory, could theoretically reach 18 percent or more. This modest six-point gain would jump the power output of each cell by half. Attaining this improvement would significantly shrink the gap between the cost of photovoltaics and that of power from new coal-burning utility plants. But it would be of little value unless the cells can be mass-produced at lower costs than prevail in the current small-scale operations. Obviously, we need a research program to reach 18-percent efficiency and to develop mass-production techniques to cut costs, just as Ford originally did with the Model T. The amorphous cells suffer from degradation over time, a problem also amenable to research. The other types of cells each have problems of cost or waste by-products, and their efficiency can likewise be improved. There are several candidates for the ultimate solar cells, and we need a technical competition, like that of the thousand rival car shops in 1900, to find some winners fast.

Even if solar cells become cheap very quickly, the advent of the commercial hydrogen car, if it depends on individual decisions by consumers, will remain remote. The fuel-distribution system is so specialized that even if other cost factors were equal, no hustling entrepreneur could ever be expected to open a hydrogen "gas station" as a profitable business. By the same token, no one could be expected to buy a hydrogen-fueled car because it couldn't get fuel to go anywhere. Because of this "network-economy" problem, there is little incentive to resolve the technical obstacles, such as boiloff, related to using hydrogen in cars.

There is a standard approach to accelerating research and development in these situations: government intervention *of the right kind*. In his 1962 paper, Arrow noted that to encourage optimal levels of innovation and invention it is necessary "for the government or some other agency not governed by profit-and-loss criteria to finance research and invention." He found several reasons why industry fails to put enough investment into R and D, which derive from the uncertain chances of success and profitability. In addition, a problem peculiar to research is that others can often copy the information produced. Although patents are a partial solution, profits from an invention are often reduced because the inventor cannot completely appropriate his own invention. For example, a new computer chip can be patented, but other firms can "invent around" the patent and, in effect, steal the idea. Taiwanese and Korean knockoffs of an IBM PC were used to write this book. Similarly, once a solution to the problem of the hydrogen tank is discovered, others can mimic it, at least in part. The investors in research face an army of eager copycats.

How can the federal government intelligently intervene to make up for these market failures? There are several options; we offer two examples that address both the network-economy problem and the issues raised by Arrow. Individual motor vehicles must be fueled by a network of readily accessible stations. But centralized fleets, such as buses, taxis, or short-haul government vehicles, can switch to alternative fuels before the larger network exists. For example, in 1897, the Electric Carriage and Wagon Co. operated a fleet of twelve electric taxis on the streets of New York. By 1900, nearly twice as many electric cars as gasoline cars plied America's urban roadways. By 1901, a recharging network was spreading from New York to Boston and south to Philadelphia. Still, recharging was frequent and slow. Although Edison and other inventors ultimately failed to solve this problem, the early urban fleets had spurred their interest even while the market remained limited.

The federal government's fleet of one-half million vehicles presents even greater opportunities. Its power to procure electric, or eventually hydrogen, vehicles would stimulate fuel-production capacity to expand, and research along with it. Later, a distribution network could grow outward from the central fueling stops. Eventually, as technical problems were solved, the average driver could be brought into the loop.

For some products, the increase in production scale following from federal procurement alone would bring costs down. Indeed, over a decade ago, proponents of solar energy within the Carter Administration argued to no avail that the production cost of photovoltaic cells would plummet if government purchased even modest numbers of modules. Why not take five federal facilities, like military bases in optimal sunfall locations, and solarize them now?

A second option is less intrusive and avoids forcing the government to make direct choices. Instead, it can issue "Requests for

Proposals'' from industry, universities, or other parties. The RFPs, as they are known, should announce that in view of the threat of global warming, there is a need to develop renewable-energy technology for power, be it for vehicles or for stationary sources. The government would guarantee to purchase a certain number of such products from those who met specific operating criteria at a particular price.

The federal government has used these techniques in the past to stimulate R and D in space and weapons technology. Some of the research produced important consumer spinoffs in the process: solar cells that were used to power early satellites now run calculators and supply electricity to homes. On the other hand, these sorts of inducements also helped to precipitate the Pentagon procurement scandal, as well as some of the horror stories associated with cost-plus contracts.

BAD MISTAKES

We can learn some lessons about such pitfalls by examining one of the past mistakes in detail, the attempt by the government to jump-start a technology—synthetic fuels—with direct handouts to industry for demonstration projects, which we mentioned in chapter 3. At the time, the estimated price of producing synthetic oil after appropriate production scale-up was about $64 a barrel. The price of drilled oil was then $35 per barrel and climbing. It was projected that the price gap would close in about ten years. The objective of government support was, as a congressional report put it, to "provide access to capital that might not be forthcoming to synfuels producers under normal business conditions." As the price gap closed, private investment could replace the public subsidy. Unfortunately, when the bottom fell out of the oil market in 1982, the gamble went sour, and synfuels could not prove economical within a reasonable time.

The Carter Administration made appalling mistakes in setting up the Synfuels Corporation. The environmental degradation associated with synfuels production would always limit acceptability in those parts of the country headed for sacrifice to the "national interest." The Administration advertised the program as the answer to an emergency, but there was no way it could replace significant amounts of imported oil for decades. To top it off, cheaper oil-saving measures, such as efficiency improvements, caused the oil market to soften and the price to drop, making synfuels even less attractive. The corporation died in 1986, having spent over a billion dollars on but four projects which produced little in return.*

If long-term alternatives to oil were desired, the Carter Administration should have bet more money on a variety of smaller, non-polluting, and lower-risk gambles. It did so to some extent, by using tax incentives to encourage some solar and efficiency investments. It also directly subsidized research that had useful results, including the development of highly efficient ballasts, the devices that maintain the electrical discharge into fluorescent bulbs. The government's Solar Energy Research Institute (SERI) supported some important advances in photovoltaics and some big boondoggles in solar thermal power. Funding subsequently dried up when the Reagan Administration slashed the Department of Energy's efficiency-research budget by half and outlays for renewable energy by 82 percent. For example, federal support for photovoltaics dropped from $160 million in 1981 to $35 million in 1988, while most of the tax credits for renewable-energy systems were terminated in 1985.

Subsidized synfuel production proved fruitless in part because of the size of the price gap that had to be closed before any market could exist. Consequently, the program turned into a white elephant, both economically and politically. Authorizing tens of billions to produce large amounts of hydrogen immediately would be a similar mistake, although investment in a modest PV-hydrogen research facility of the sort being built in West Germany would be prudent. But some renewables already have a significant market, which could gradually expand as research produces technical advances. Government procurement of photovoltaics, for instance,

---

*The amount spent was only a small fraction of the grandiose plan laid out in 1980, when $88 billion was authorized by Congress and $18 billion was appropriated. Counting subsidies dispersed by other agencies, the government spent only about $3 billion on synfuels during the 1980s. This expenditure was still six times the total federal support for R and D of photovoltaics over the same period.

would permit a growing business to burgeon. In contrast, the oil flowing from the one Synfuels Corporation project still supplying any, costs twice the market price to produce, so no one would buy it without a huge subsidy. There never was a preexisting market to expand and none developed. The same might be said of commercial nuclear power reactors, because none has been ordered in the United States for more than a decade.

Procurement and direct subsidies are but two of many ways to spur R and D. There are other approaches, to be explored in greater depth in the next chapter, which seek to encourage development of new technologies by discouraging old ones, particularly through economic incentives. As a first step in this direction, the playing field between fossil fuels and renewable energy should be made level by ending favoritism for fossil fuels through tax preferences such as depletion allowances. If fossil fuels become more expensive, research on renewables that might be substituted for them becomes more attractive.

Another reason these fuels are cheap is that the consequences of burning them escape market-pricing mechanisms. For example, if a factory dumps its hazardous waste on your lawn instead of paying to have it carted away, you can sue the company for the damages done. As a result, the company could pay the judgment and continue to dump; or pay a carter (presumably a bit less) to haul away the waste. Either way, the cost goes into the price of the product. But you can't sue anyone for emitting carbon dioxide and making the world warmer. As a result, the damage done to you by global warming, or the cost of avoiding it, is not included in the price of the fuel. Environmental costs of using fossil fuels cannot be fully quantified, but they could be taken into account in part by slapping a tax on the price of gasoline, for instance.

This more generalized approach to encouraging research would guard against excessive government involvement in specific technological choices. As an economist might say, the absence of pressure from an ex post market test tends to make a mockery of the ex ante marketing judgment process. In simple English, this means that you don't want bureaucrats picking winners and losers from among several technologies. Even the record of Japan's legendary Ministry of International Trade and Industry (MITI) is questionable with regard to the particular choices it has made. Japan's success at R and D proceeds more from MITI's encouragement of cooperative behavior within the economy than from its specific decisions.*

Concerns about bureaucratic choices notwithstanding, betting on a few particularly promising technologies, such as picking photovoltaics from among several solar options, would occasionally make sense. At the same time, government should probably stay away from micromanaging technologies, for example, choosing amorphous over single-crystal cells.

While government isn't exactly the mother of invention, it certainly can act as the midwife. In this case it must, because without new taxes fossil-fuel prices may stay lower than renewable-energy prices, under most circumstances, for a long time, even with technological advances. Then solar power would continue to expand slowly and only into specialized niches, such as meeting peak power demand at optimal locations. Industry will see little profit opportunity in a hydrogen system for many, many years unless government signals a developing preference for alternatives to fossil fuels and indicates that it will act to make the latter more expensive.

If these arguments on federal intervention in the R-and-D sector seem tiresome and familiar, it is because few American politicians speak without the buzzword "competitiveness" tripping from their lips like a mantra, no matter the topic at hand. This cacophony of cant arises from the loss of consumer-product markets by U.S. companies, and the resulting balance of payments deficit, and the failure of companies to invest in R and D, which is one source of the problem. Even before President Bush took office, his economic team called—you guessed it!—the Competitiveness Study Group was batting back and forth ideas aimed at encouraging these sorts of long-

*Japanese companies are encouraged to launch joint operations at the product-research level and later split apart to compete in development and marketing. MITI also supports R and D centers that diffuse technology through the economy by transferring it to small- and medium-sized firms. With the support of the Department of Defense, the United States is now experimenting with similar approaches, such as the Sematech Corporation, a consortium of companies developing manufacturing techniques for advanced computer-chip technologies. The Pentagon is also supporting a modest joint-venture research venture on high-definition television, an arena in which the Japanese apparently have a head start.

term investments by industry. Still, the message is no less valid for having been overused. After years of complacency, politicians are finally awakening to the realization that a wave is breaking and that America will either rise on it or be drowned.

THE NEXT WAVE

It is impossible to consider the solution to global warming without examining this broader economic context, and it is impossible to understand our economic fate without considering the havoc that will follow from disregarding environmental constraints. Significant warming would destroy much of the natural-resource base underlying the global economy. The inverse is also true: the solution to the greenhouse problem will involve the utilization of new forms of energy technology, and energy transitions have been part and parcel to spurts of economic growth in the past.

For the two centuries since the advent of the Industrial Revolution, economic activity in the capitalist nations has moved through periods of expansion and stagnation. In the 1930s, Harvard's Joseph Schumpeter suggested that long-term economic cycles might be related to the process of technological change. The advent of apparently endless expansion and a strong hand in fiscal and monetary regulation by government ended interest altogether in this notion. But the slowing of the global economy following OPEC's oil squeeze has caused some economists to reconsider Schumpeter's ideas.*

*In the 1920s, a Russian economist, Nikolai Kondratiev, first wrote of a forty- to sixty-year cycle in an article, "The Long Waves in Economic Life," which was published in English in 1935. By then, Kondratiev had disappeared forever into the gulag after his work led him into ideological disputes with Trotsky, and was reviewed in the 1929 edition of the *Great Soviet Encyclopedia* with this one sentence: "This theory is wrong and reactionary"—presumably because it allowed for the continual rejuvenation of capitalism. Schumpeter took an interest in this work, but Kondratiev's methods were later shown to be faulty. Still, some recent disciples of Schumpeter, such as Christopher Freeman and Carlota Perez, use the term "Kondratiev wave" to connote the long-term cycles that they find to have characterized economic growth and technological change.

Simply put, Schumpeter and his recent disciples have argued that invention and innovation spur waves of investment, economic growth, and technological development by releasing the "animal spirits" of entrepreneurs. Essentially, one such wave occurred about the turn of the century in the United States as the automobile and other industries grew explosively. After some period of time, the rich vein of progress resulting from a spate of invention is mined, investments produce declining returns, and the psychological climate for new investment becomes exhausted. Then the economy stumbles, as it did in the Great Depression of the 1930s. Technological progress has a wavelike pattern because invention and innovation are spaced unevenly, occurring in spurts.

The first industrial wave, between the 1780s and 1830s, received its impetus from water power, canal transport, and textile manufacture. The second, between the 1830s and 1890s, was driven by steam power from wood and then coal combustion, along with railroads and machine tools; the third ran through the Great Depression and saw the growth of electric power, steel fabrication, synthetic chemicals, and radio, using standardized production. The fourth has seen the rise of oil, automobile and air travel, the peak of mass production, petrochemicals, and military–industrial production. An incipient fifth wave is based on information technology, such as computers and fiber optics, and is characterized by efficient energy use; robotized, decentralized, and customized production; and information services.

Each wave entails not just a particular set of innovations but an economic and social structure that reinforces, and is reinforced by, the new innovations. This technical–social–political framework has been referred to as a paradigm.* For instance, the fourth-wave postwar boom was predicated on cheap energy, particularly oil for transportation. The boom was reinforced by the federal government's subsidy of housing construction, the massive expansion of the inter-

*In this context, the use of "paradigm," which means "pattern," was borrowed by economists Christopher Freeman and Carlota Perez from Thomas Kuhn's *The Structure of Scientific Revolutions* (Chicago: University of Chicago Press, 1962), where it was originally used to indicate the set of problems, solutions, and assumptions that govern science at any one time. Here it means an entire technological, social, economic, and political approach to solving the problem of production.

state highway system in the 1950s and 1960s, and the advent of the jet engine in commercial aviation, which gave rise to an entire peripatetic culture. Superhighways and fast airplanes spread the scope of markets for goods until they became global. These opportunities in turn prompted innovations in mass marketing (as late as 1946 there were only eight shopping centers in the entire country), the creation of suburbia as a place in which to work as well as live (after New York City and Chicago, Fairfield County in Connecticut is the third leading corporate-headquarters site in the United States), and new forms of industrial organization (multinational corporations).

But development of the interstate highway system, jet engines, and mass markets depends on factors (such as war and its aftermath) that are not necessarily timed with technical innovations because they are also caused by a variety of other forces. If the economic–social–political system is mismatched with the technological opportunities, the boom may be stillborn. For instance, the internal-combustion engine was perfected in 1876 by Nikolaus Otto in Germany, but it had little economic impact until it was mated with Ford's industrial system, and the burgeoning U.S. economy, in the early 1900s. Not until society catches up with the technological possibilities will the changeover to a new economic paradigm occur.

There is an analogy between this kind of paradigm shift and the transition that occurs when water freezes. Just as two economies with access to the same technical information and resources can have vastly different productivity, water at thirty-two degrees Fahrenheit may take three entirely different forms, either liquid, solid, or gaseous. Liquid water molecules are loosely organized and somewhat randomly oriented, which is why water flows. Still, at thirty-two degrees the spatial relations and the forces between H_2O molecules can become reorganized. The attractive force of molecule A for its neighbor B tends to lock both into the particular alignment, which we will call the "heads-up" position. But B will also tend to lock into "heads-up" with its neighbor C, which tends to lock with its neighbor D, and so on. This coherence of force and position propagates suddenly through an entire chain of H_2O molecules and out in all directions, rapidly forming a crystal of ice, with all molecules locked in a rigid configuration with the heads-up orientation. A phase transition has occurred, characterized by a structural reorganization.

Another example is the emergence of a mogul or bump on a ski slope. All routes are equivalent on a fresh slope. But after a skier randomly picks one particular route, it becomes easier for others to follow it, so hills and ruts develop. Ski tracks attract other skiers, just as one molecule attracts others to its initial random orientation. Similarly, one invention may spark a spurt of others and lead to an entire technological transition. Technological and social-cultural factors (education, infrastructure, industrial organization) also cohere in a mutually reinforcing manner,* just as spatial relations and force do in a liquid. The system is locked into one state like a block of ice. To make a transition to another state, considerable reorganization must occur because no one element can be punched out and reoriented in isolation. Consequently, two economies with access to nearly the same sets of inventions or innovations might organize themselves in totally different fashions, like water at thirty-two degrees.

The paradigm concept is, of course, imperfect, but some economists have found this model to be useful. Christopher Freeman and Carlota Perez of the University of Sussex in England have employed it to describe not only historical experience but also the current peculiar state of the U.S. economy. They argue that this country finds itself mired in the fourth wave, while Japan is much farther along in transition to the fifth.

As we have seen, key technological advances in energy utilization have been associated with each new wave. First, water power, then direct use of steam power, then electrical power, and finally oil power in transportation characterized successive technological swings. Shifts in the primary energy source can occur at a furious pace. In 1870, 75 percent of U.S. combustion energy came from wood, 25

---

*For example, once families moved to suburbia and needed to buy cars for some purposes, other transportation options were driven out for all other uses. High energy use generated a living pattern which led to more of the same. A dispersed pattern of living precludes efficient use of energy in a variety of ways. For example, residences can be heated and provided with electrical power simultaneously through cogeneration from central power plants. This process, called "district heating" (discussed in chapter 6) is only feasible when housing patterns are concentrated.

percent from fossil fuel, almost entirely coal. Only thirty years later, the percentages had reversed. The shift to oil was more gradual but still impressive. Between 1945 and 1970, coal fell from 50 percent to 20 percent of U.S. fuel consumption as oil and gas use soared.

So far, the fifth wave has not seen a substantial shift in energy sources. But compared to the sixty-year time frame to commercialize technologies that we noted earlier, silicon photovoltaics are still young, originating only in 1954. But already the new information-age production methods allow more efficient energy utilization through computerized control systems, so that low energy intensity is becoming embedded in the structure of the economy in places like Japan (which was more vulnerable to the 1970s oil-price increases for lack of domestic supplies and therefore moved quickly to reduce its dependence on foreign oil) and to some extent, the United States, too.

Most of the savings initiated with OPEC's oil-price hikes in 1973 and 1979 have remained even as the cost of oil has returned to pre-1973 levels.* The reason is partly that some investments have a long life, partly that new habits were learned and innovations in production were developed, and partly that changing political factors generated regulations such as fuel-economy standards for automobiles in the United States.

The extent to which efficiencies have been retained, that could easily have been reversed once oil prices dropped, is a reflection of structural reorganization precipitated by, but not dependent on, high oil prices. Yet although the United States has made significant progress in reducing energy inefficiency, it still uses twice as much energy as Japan or West Germany per unit of gross national product, and large swatches of its fourth-wave economy remain in place. As long as this is the case, the United States cannot take full advantage of available technologies, and its competitiveness may lag.†

---

*The amount of energy used to produce a unit of gross national product in Japan decreased by one-third between 1973 and 1985; in the United States it decreased by more than one-quarter.
†A report by D. P. Levin ("New Japan Car War Weapon: 'A Little Engine That Could,'" *New York Times,* Nov. 26, 1989, p. 1) provides an interesting example. Concern for fuel economy and leadership in precision manufacturing has given Japan's auto industry a head start in the competition to produce the high-fuel-economy, high-power multivalve engine.

What has all this economics to do with climatic change? Global warming is a leftover from the old fourth-wave paradigm. The world of the fifth wave will be an energy-efficient one, utilizing computer controls in production and lightweight yet strong ceramic and composite materials in transportation. Home computers linked by fiber optics to the workplace will reduce the need to commute. The main energy source could be solar cells manufactured from the same raw materials as computer chips. Electricity will displace combustion in factories; power generation will be modular and partly decentralized, using photovoltaics and fuel cells in order to maintain flexibility in attempting to match changing, customized demand requirements. As the fifth wave unfolds, global warming could become a thing of the past. *The type of innovations that bring on the fifth wave are the same as those that will stop global warming.*

Sooner or later, unrestrained global warming would in any event force a structural reorganization that necessitates a full transition to the fifth wave. After all, the transition between paradigms has always proceeded in part from the limitations of the earlier one. But the inevitability of change cannot be a cause for complacency in this case, because once the consequences of warming are so manifest that they become real constraints on economies, the world will be committed to potentially catastrophic and irreversible climatic changes down the road. Environmental collapse would become the agent of industrial transition. Needless to say, there ought to be a better way to approach the inevitable.

From this perspective, it is unrealistic to believe that the United States will lead the way out of the greenhouse quandary unless it solves its current economic problems; and it is worth remembering that the transition to the fourth wave didn't come easily either. World War II and the accompanying economic and social upheavals provided the impetus for the last transformation. Now, those patterns must be shifted again.

The transition out of the fourth wave depends on our renovating the system of education, worker training, and research. In particular, it will mean escaping the military–industrial paradigm and entering an environmental–industrial paradigm, because the postwar value of military research to civilian industries has been exhausted. Nearly two-thirds of federal R and D spending and over one-third of *all* R and D dollars (public and private) go to military purposes. As Columbia University economist Richard Nelson puts it, "Since 1972, it is arguably the case that military R&D has cost the U.S. considerably in terms of foregone civilian alternatives." It is ironic that Japan lagged in civilian mass-production technology before World War II due to an emphasis on military applications. The tables have now turned.

Even the military now realizes that the impoverishment of civilian research and development undercuts its own agenda. Since 1988, it has been charged with developing a yearly "Critical Technologies Plan" designed to guide congressional decisions on support for technologies that might turn out to be useful in weapons systems. Some of the twenty-two listed in the first group—for instance, methods for preparing gallium arsenide, a component of some solar cells—might help slow global warming. If so, the connection would be fortuitous, an accidental spinoff.

The Defense Department has carried the notion of technology boosting even farther: it has proposed a new system of economy-wide technology support run out of the Pentagon, a sort of techno-octopus, which would invade every sector of technological life. (This poorly timed suggestion came just as the 1988 procurement scandal was breaking.) For instance, the Pentagon would encourage specific education and training programs that could muster a work force keyed to the development of high-tech weapons.

Aside from Orwellian "1984" worries conjured by this scheme, the Pentagon's usual clients are not necessarily big innovators anymore. As Polaroid's executive vice-president Sheldon A. Buckler noted, "Flash of genius is a wonderful thing when it happens, but it often doesn't happen more than once in the history of a company." Still, the Pentagon got one thing right: the fifth-wave paradigm involves not just a reordering of federal expenditures but a reordering of the very structure of research and more. The United States may need a MITI or it may not; it may or may not need Japanese-style consortiums; but it does need some kind of a structure to provide government leadership.

The atmosphere is becoming what oil became for some nations after 1973, a resource used to the limit, a constraining factor on further development within the same paradigm. By making atmospheric preservation a principle to guide technological development, the path to the fifth wave will be shortened. Should there be doubts about the efficacy of government involvement, consider that to the extent that World War II accelerated change, it was a politically sponsored transition. To believe that a less malevolent, more intentional approach is bound to fail is too cynical a view, considering the stakes involved. Either the United States will accelerate the global transition by pressing industrial reorganization or we will continue to lag behind. In that case, the best we could hope for is to wind up buying photovoltaics stamped "Made in Japan." The worst is that our inertia will so hamper the global transition that no nation will have much future at all.

CHAPTER 10

Building Blocks:

A Plan of Action

T HE FIFTH WAVE IS A VISION OF THE FUTURE; IT WILL TAKE decades to establish a new paradigm firmly. Nearly half a century could pass before the hydrogen economy matures, even with governmental intervention. What about the here and now? There is simply no time to wait for research and development to pay off.

As we have seen, the efficiency bridge of chapter 6 is at the same time a route to the fifth wave and an interim means to slow global warming. Incentives to use less fossil fuel do double duty: they cut emissions while they spark more research on fifth-wave solutions (just as impending restriction of CFCs led to rapid development of substitutes). Slowing warming in this way will make adjustment easier and reduce the likelihood of a climatic surprise analogous to the ozone hole. Such a shock would engender ill-conceived political responses, in the way that the 1979 oil-price shock led to the Synfuels Corporation. But what specific steps should be taken and what are the politics of implementing them?

The availability of a cornucopia of cheap efficiency measures does not assure implementation of a single one. As we saw earlier, ostensibly cost-effective devices, like compact fluorescent bulbs, are routinely ignored by manufacturers, distributors, and consumers, due to differences in quality (which exacts a price), lack of information (which is easily but not freely overcome), and unwillingness to bear large front-end costs. Much more will be required of government than to encourage development of additional technologies if greenhouse gases are to be reduced.

Government may use several forms of regulation to make everyone in the marketplace snap to at once. Economic incentives such as taxes make one thing more expensive than another in order to attach a penalty to its use. Alternatively, so-called command and control aims to change behavior by direct prohibition of specific activities on the part of particular producers or consumers, and it usually winds up costing people money, too. For example, gasoline taxes provide an economic incentive for efficiency, while specifying a particular level of fuel economy for each automobile is a command-and-control approach.

Setting a tax at a rational and effective level proves difficult in practice. For instance, the amount of emissions reduction brought by a given level of gasoline tax is difficult to predict in advance. Attempts to use an alternative approach, such as tying taxation to the costs of environmental damage caused by gasoline combustion, suffer from the lack of knowledge of the extent of damages, particularly for global problems. For these reasons, there is no straightforward way to incorporate specific goals, such as a 50-percent emissions cut, into a tax-based system. Command and control can impose specific environmental targets like a 50-percent emissions cut by every single source, but the drawback of this approach is that it tends to be unnecessarily expensive.

Just as a less invasive hand by government might produce more effective R and D, a "light-hand" approach could avoid some of the problems of command and control while returning the same benefits. A third option, called a permit system, has recently been advanced by economists like the Kennedy School's Robert Stavins and was incorporated into the Bush administration's 1989 acid-rain proposal. It combines the economic incentive characteristics of taxes with the command-and-control advantage of targets. In the case of sulfur-dioxide emissions reductions, for example, a target like a 50-percent cut is chosen, and paper obligations to reduce emissions are assigned to each power plant owner. But obligations can be traded among owners. If one owner thinks he can cheaply make reductions beyond his own obligation (for instance, by construction of a non-fossil-fuel plant), he lets a second plant owner pay him to relieve the second owner of *his* obligation. Of course, the latter owner will only do so if the payment is less than what meeting the obligation at his plant will cost.

This sort of pollution control is particularly appropriate for dealing with greenhouse gases. For example, if the U.S. utility industry were told to cut carbon-dioxide emissions by three-quarters, it would be far cheaper to allow companies in the sunny half of the country to go totally solar and let northern companies continue to use some coal than flatly to cut emissions at every company or plant in half. In toto, the same carbon-dioxide cut would be obtained.*

GETTING STARTED

The politics of none of these approaches is simple, and a combination of all three will be needed to deal with global warming. Fortunately, a political path of least resistance has developed which will allow us to begin. Fossil fuels are already at issue both here and in Europe, as decisions move forward to reduce smog and acid rain. Cutting fossil-fuel use by encouraging efficiency will generally cut smog and acid rain, as well as carbon dioxide and several other greenhouse gases. Some of the measures proposed below are designed to accomplish the triple task of controlling smog, acid rain, and greenhouse gases in one fell swoop, in this way making the overall effort cheaper; and these measures require doing political battle only once, instead of three separate times.

Motor-Vehicle Fuel-Economy Regulation

Greater fuel efficiency in motor vehicles means lower carbon-dioxide emissions. Depending on how it is achieved, fuel economy

---

*The permit system has another advantage: it encourages innovation because technologies that yield larger-than-average emissions reductions produce something that may be sold. Under command and control, reductions beyond the mandated level are worthless to a plant operator.

could also reduce emissions that contribute to smog and acid rain. Current U.S. Corporate Average Fuel Economy (CAFE) standards have contributed to an improvement in the efficiency of new cars from fourteen miles per gallon in 1973 before regulations were legislated, to twenty-eight miles per gallon in the late 1980s, when the Reagan Administration began deferring further progress. A goal of fifty miles per gallon should be set for the year 2000. Regulation could employ command and control enforced on each manufacturer's fleet, or a permit system allowing trading of obligations among manufacturers.

Currently, highly efficient cars save money in the long term but generally lack Cadillac comfort. Future efficiency gains through engine and transmission redesign and the use of lightweight materials instead of cutting size would be welcomed by the public. If fuel economy is tightened simultaneously with air-pollution limits, auto manufacturers would be coaxed toward vehicle redesign to optimize both.

Regulating Electric Power

Great efficiency improvements are made, not born. For example, state efficiency standards for appliances and federal requirements for informing consumers have helped spark a doubling of refrigerators' electricity efficiency over fifteen years. New federal standards for appliances will be promulgated shortly. With R and D, we should be able to upgrade efficiency continually and tighten the regulations.

Power companies appear every year or two, hat in hand, before public utility commissions in each state to have their rate structure reviewed. Generosity on the part of state regulators should be predicated on the delivery of lowest-cost, low-emission energy, including services that provide it in the form of efficiency improvements (for example, have the power company agree to supply your $15 compact fluorescent lightbulbs and spread out the upfront cost in electricity rates over time, as the Southern California Edison Company has done).

Electric utility companies provide a convenient avenue for overcoming the "household discount" problem, the consumers' reluctance to bear higher purchase costs even when greater long-term savings would accrue from lower electricity use. According to physicist Arthur Rosenfeld of the Lawrence Berkeley Laboratory, customers won't buy efficient appliances unless the payback occurs in less than three years—equivalent to a whopping 40 percent return on his investment! For some purchases, consumers won't budge without a six-month payback on the extra cost. In contrast, utilities have long planning horizons and demand much lower returns, about 15 percent, so they could make the initial purchase instead of the consumer and then spread the cost over a number of years in the rate base. Utilities in California and the Northeast are exploiting such approaches, while most other states are still asleep at the wheel.

New York State has just instituted a novel program that requires utilities to consider environmental costs explicitly when investing in particular generation alternatives. A sliding scale has been developed to account for those costs by favoring efficiency with a 15-percent credit over coal, for example, when a utility considers the cost of each new supply option. Suppose that a utility needs 1,000 megawatts of power to satisfy new customers; it might build a new coal plant or it might equally well pay several of its industrial customers to install efficient motors in their factories. Say the new plant costs $1 billion and the motors $1.1 billion. Previously, the coal plant would have been selected. But with 15-percent added to the price of the coal plant to reflect environmental costs, the efficiency investment becomes cheaper. Fifteen percent may not fully capture the environmental costs of coal, but it's a good beginning.

These proposals are politically simple and save money to boot. There will be new opportunities to employ such strategies when the states are presented with the task of implementing federal acid-rain legislation. Utilities should be required to consider efficiency on an equal footing with other emissions-control strategies and to use the sorts of consumer incentives noted above to implement it. Scrubbers cut sulfur dioxide at $600 per ton, while an array of energy-efficiency measures can get part of the job done at one-third the cost.

Fuel-Use Taxes

Call it a fee or a tax; a levy slapped onto the price of fossil fuels will discourage combustion and encourage efficiency and renewable energy. Regulation and taxes go together like hand and glove: for instance, CAFE regulations compel manufacturers to produce efficient cars, and fuel taxes encourage the consumers to give up their current gas-guzzlers more readily. That's one reason why Lee Iacocca, chairman of Chrysler Corporation, which already leans toward efficient models, strongly supports a higher gasoline tax.

Yet not knowing the benefits of any given level of taxes is particularly nettlesome in this case. Although new-car fuel economy doubled as gasoline prices rose in the past, CAFE standards were simultaneously on the increase and a national attitude of energy efficiency was being encouraged, at least initially. Demand for gasoline may be highly inelastic, for example, requiring a doubling of the price to dent the market. In Europe, gasoline prices are double U.S. levels, while cars are merely 20 percent more efficient. Would the public sit still for a doubling of gasoline prices? Might it accept a tougher CAFE standard more readily, or a little bit of both, gradually? We are not just experimenting on the atmosphere here: we need a quick testing of the political system, too.

The tax-and-spend image has driven Democratic presidential candidates into oblivion for a decade, while George Bush and the Republicans have taken an anti-tax blood oath. The public seems to prefer implicit taxation through regulation of industry to measures that bite them directly, so improved CAFE standards may be the place to start. And while we're at it, consideration should be given to developing a new credit system for car buyers analogous to the rate-basing scheme used in California to pay for efficiency measures. This would decrease the resistance to more costly but highly efficient vehicles. The two problems are the same.

Like a gasoline tax, a surcharge could be placed on coal and natural gas purchases, weighted for the amount of greenhouse gas each produces, in order to discourage all fossil-fuel combustion. Similarly, tax write-offs and rebates for clean energy could provide balance in a skewed system, for the federal government already gives renewable energy a comparatively hard time by favoring fossil fuels and nuclear power with various subsidies and incentives. For example, a gas-guzzler fee on cars with low fuel economy could provide funds for rebates to purchasers of cars with high fuel economy. A legislative package, including incentives to industry, like procurement, which we discussed in chapter 9, along with the incentives to consumers noted here, would put us well on our way toward solving the fossil-fuel dilemma.

An End to CFCs

Chlorofluorocarbons have got to go, in order to protect the ozone layer and to stabilize the climate. The Montreal Protocol process must be continued until substitutes are firmly in place, preferably by 1995, at the latest by the year 2000. Care must be taken to ensure that ozone-friendly substitutes are also greenhouse-benign.

More Mass Transit

In many metropolitan areas, commuting is a disaster. Carpools, bus lanes, and commuter vans can help, but fully utilized rail mass-transit systems are a critical part of any attempt to reduce smog and greenhouse-gas emissions. After a decade of assault from critics, the benefits of mass transit remain clear: well-loaded trains and buses save energy and cut pollution, and such operations are feasible in many areas now. With improved metropolitan design, more regions could usefully avail themselves of mass transit (see chapter 7).

Forty-one mass-transit systems are under construction in the United States, largely due to earlier federal largess. But the fiscal pie has shrunk. If experience with sewage-treatment plants is any guide, these systems may never expand to their full potential because of inadequate support from all levels of government. Instead of allowing these new systems to fail, a larger chunk of the federal gasoline tax should be allocated to them.

Adequate mass transit and environmentally benign regional planning present another chicken-and-egg dilemma: each is impossible without the other, and each makes the other easier to achieve. To move the process along, states should be required to develop comprehensive transportation plans that limit air pollution and greenhouse gases before they become eligible to receive federal transportation grants. Economic incentives to the individual are also badly needed. For instance, drivers in Stockholm are charged a fee of $45 per month—but the same permit allows free access to mass transit. With population piling up in two coastal corridors (over half of Americans now live within fifty miles of the sea), mass transit in the United States is becoming more feasible even for intercity travel.

Plant a Tree

An inexpensive way to slow warming while politicians gather courage for regulation is to plant trees. The Environmental Defense Fund's economist Daniel J. Dudek has designed a program to offset the effects of carbon-dioxide emissions from future additions to the U.S. power-plant inventory. The Agriculture Department's Conservation Reserve pays farmers to take erodible land out of production and plant it with trees. Contributions to this program by utilities could build an additional ten-million-acre carbon bank planted in fast-growing silver maples at a cost of $1 to $2 billion. True, the trees would not affect levels of other pollutants much, but the program could be as cost effective as carbon-dioxide reduction through implementation of efficiency, so tree planting provides an economical compliment to such efforts. For example, in early 1989 the City of Los Angeles announced a tree-planting program to provide energy-saving shade for houses in order to reduce air-conditioning needs. People love trees, so tree planting is politically attractive, cheap, and environmentally beneficial.

Other Gases

Although the relative importance of each source of the greenhouse gases other than carbon dioxide and CFCs is unknown, opportunities abound to reduce every one of them. Methane and nitrous oxide* may be largely by-products of the warming itself. But it would be prudent to assume otherwise and to adopt some easy measures to hedge the risk.

*Increasing temperature may accelerate the processes in soils that produce nitrous oxide, just as George Woodwell has argued for methane.

Methane from solid-waste (garbage) landfills can be tapped and burned, replacing fossil fuel and keeping the methane out of the atmosphere. If coal burning is thus avoided, both carbon dioxide and methane levels are cut. Coal mines, oil-drilling operations, and pipelines can be tightened to reduce methane leakage. Carbon-monoxide emissions from cars can be cut to preserve the atmospheric reservoir of the beneficial chemical hydroxyl, which keeps methane in check. Research on new rice strains and cattle-feeding regimes must be expedited.

More research support is also the key to slashing nitrous-oxide emissions. We need to know precisely how much comes out of power-plant stacks and how to reduce it. Similarly, alternatives to the particular chemical fertilizers that generate nitrous oxide from soils must be developed, along with procedures for keeping sewage and fertilizer nitrate out of rivers and estuaries, as some of it ends up as nitrous oxide in the air.

Third-World Rescue

While much of the Third World is drowning in debt, it is foolish to expect these countries to implement long-term environmental initiatives. The fifth wave in the developed nations may pull up the Third World, but in the process developed nations may merely shift their polluting production to emerging nations, by which time forests will have been destroyed to pay foreign debt in any case.

There are four immediate remedies we should adopt. The United States should lead the way in debt reduction, permitting internal growth to replace external payoffs. Transfer of renewable-energy technology should be arranged so as to encourage local production, research, and development, instead of indefinite dependence. We should assist family-planning efforts where there is local demand and support measures that enhance the status of women so that they can take advantage of such programs. And where political change, land reform, and economic redistribution would favor greenhouse-gas reduction, such as in Brazil, First-World aid and lending policies should urge it out of pure self-interest.

More Research and Development

Until regulation forces reduction in greenhouse-gas emissions, research on hydrogen cars or low-flatulence cattle promises little profit for industry. If a problem has a potential solution that will earn no profit, then government must, one way or the other, pay for the research involved. The federal R-and-D funding process needs renovation, primarily by transferring resources away from military end-use to the energy–greenhouse-gas problem. For 1990, the Pentagon's proposed R-and-D budget was $44 billion, while all federal R and D on the global environment amounted to $191 million. Even small shifts toward addressing the problems of global warming could make a big difference.

We cannot overestimate the importance of research on both the climate problem and its solution. Between now and the time that the political climate permits stringent regulation of greenhouse gases, industry will not be inclined to spend resources on supposedly chancy technologies, such as hydrogen. By the time public fears demand regulation, the planet may have fallen over the three-degree cliff. We need a commitment comparable to the Manhattan Project in order to analyze the problems ahead and produce solutions. A research program fostered by government can assure that answers are available immediately once the public has become primed to apply them.

In the meantime, government should try to avoid what Jessica Tuchman Mathews of the World Resources Institute calls the "regret factor," that is, any decision that inadvertently makes things worse. Recent cuts in Amtrak and mass-transit funding are a step in the wrong direction, and a large-scale commitment to methanol use in cars would be as well.

Start Talking

In November 1988, under the auspices of the United Nations Environment Programme and the World Meteorological Organization, more than sixty nations established a scientific and policy-assessment process in Geneva called the Intergovernmental Panel on Climate Change (IPCC), similar to the one set up in 1980 to deal with ozone depletion. With governments still shaping their attitudes about global warming, just sitting down together and talking over options might produce movement toward limiting emissions as it did with the problem of ozone depletion.

In addressing the opening session of the IPCC working group on policy options in January 1989, U.S. Secretary of State James Baker said, "We can probably not afford to wait until all the uncertainties have been resolved before we do act. . . . Time will not make the problem go away." Unfortunately, holdover Reaganauts, some of the very same people who were part of the "hats and sunglasses" crowd during the ozone debate, still had a choke hold on the Bush administration's global-warming ruminations, and they proceeded to undercut Baker with endless delaying tactics, proposing nothing but more studies.

By the next climate-change meeting in May 1989, the situation had deteriorated even more because Bush's chief of staff, John H. Sununu, had decided to reject the call by several European countries for immediate negotiations on a convention. But at the same time, the Office of Management and Budget was caught trying to alter James Hansen's congressional testimony on the seriousness of global warming. The ensuing flap gave EPA Administrator William K. Reilly the chance to get Sununu's decision reversed. But the administration still has no coherent policy on the greenhouse effect. If we dare to assume that the United States will wake up, the Second World Climate Conference in 1990 could provide a venue for the first steps toward a global climate accord. If the Bush administration slumbers on, hopefully others will grab leadership on this issue.*

Après Moi, le Déluge

What is true of U.S. industry holds true for the rest of the developed world. Automobiles may attain slightly higher fuel economy

---

*The possibility remains that a policy may yet emerge. MO briefed William K. Reilly just before he was sworn in as EPA administrator, and his understanding and concern were obvious. Similarly, the chairman of the Council on Environmental Quality, Michael Deland, has shown a strong interest. Unfortunately, the May 1989

elsewhere, but manufacturers abroad could certainly do much better. The same applies to electricity efficiency. The politics of legislating even cheap measures is not simple anywhere; without political leadership, it is impossible.

While he was still at the Environmental Defense Fund, attorney Joseph Goffman, now a Senate staff member, pointed out that no environmental legislation ever passes the Congress just because the costs of damage exceed the costs of avoiding it. The fact that air pollution and acid rain cause adverse effects on human health, resulting in lost workdays because people suffer more respiratory diseases, and that this ongoing cost exceeds the price of emissions controls; that air pollution and acid rain kill life in vulnerable lakes, streams, estuaries, and forests, natural-resource bases that may take centuries to recover; that they cause the corrosion of bridges, buildings, and monuments; and that they hasten the destruction of books and works of art that are the highest expressions of human thought and creativity—all of this counts for naught to most members of Congress, who pantingly ask the really *big* question: "*What's in it for my district?*" In the real world, the key to solving big problems is that each individual constituency must see its own particular advantage in the larger solution.

Informed, intelligent leadership is necessary to organize such coalitions, and there has been little leadership on the environment from U.S. politicians. Numbed by years of propaganda against federal intrusion, they often lack the wit and courage to proffer solutions.

In Europe, at least, a new day has dawned. The growing strength of environmental parties is forcing even conservative politicians to raise the green flag. Chernobyl and the Rhine River chemical spill generated public anger, while the success of the Montreal Protocol

episode was replayed in November 1989 when, at a ministerial meeting in the Netherlands, the U.S. balked at agreeing to a goal of stabilizing emissions by the year 2000. Under pressure from the public, most of the European Community and several dozen other countries, the Bush Administration then shifted gears and agreed to a stabilization "as soon as possible" but with no date specified. MO briefed Science Advisor D. Allan Bromley shortly thereafter; Bromley's attitude, which appeared to be flexible, may be the key to developing a forward-looking policy.

on CFCs has created confidence in the possibilities for global co-operation. A new planetary paradigm is in the making, and it has geopolitical as well as economic aspects. Gorbachev has given the West a chance to rethink its priorities. Technological progress may finally detach from military exigencies and become firmly soldered to environmental needs.

On November 17, 1988, one year before the Berlin Wall tumbled, British Prime Minister Margaret Thatcher said, "Assuming that Mr. Gorbachev's reforms come into effect, dare I say the Cold War is already at an end." On April 26, 1989, she convened an extraordinary briefing by scientists at 10 Downing Street for herself and selected members of her cabinet, including Foreign Secretary Sir Geoffrey Howe.* Global warming was the only subject to be discussed. Despite a developing crisis within NATO over short-range missiles, Mrs. Thatcher listened avidly, asked questions, and engaged in vigorous give and take on the subject for six and a half hours. She wasn't once distracted by other business. Two weeks later, her UN ambassador, Sir Crispin Tickell, placed a proposal for a global climate convention before the world body. Mrs. Thatcher had surely noticed the Green tide that has been lapping at West German Prime Minister Helmut Kohl's feet, and in holding the briefing on global warming she was also launching a preemptive strike on British Greens just before European Parliamentary elections.† Whatever her motives, though, her interest in global warming appeared sincere. Has any U.S. president ever devoted six and a half hours straight to the global environment?

Whether Thatcher's concern, and that expressed by other world

*Howe has since been replaced. This may cause little change in British policy on global warming. After all, MO participated in the briefing and noticed that Howe, in contrast to the Prime Minister, slumbered through much of it.

†The NATO crisis mentioned earlier was itself precipitated by the so-called Red–Green alliance of environmentalists and Social Democrats who opposed modernization of the existing short-range missile force in West German territory. It forced a confrontation and an eventual compromise between Kohl and President Bush. Subsequent to the Thatcher briefing, British Greens garnered 15 percent of the vote in European Parliamentary elections.

leaders, will translate into action remains to be seen. When even painless measures are avoided, Draconian ones, such as rationing fossil fuels, become inevitable. If Thatcher, Bush, Gorbachev, and others do not soon start building the bridge to the fifth wave, they will bequeath an impossible task to their successors. Or, as Louis XIV put it: "Après moi, le déluge."

CHAPTER 11

The Last War:

Beating Swords into Solar Cells

GLOBAL WARMING MEANS RAPID CHANGE, AND RAPID CHANGE means the end of familiar places. It means the trees that we walk among, that our forebears walked among, that we wanted our children to know, will vanish. It means that the creatures of Yellowstone will scatter and the languid salt marsh, with its clams and ducks and egrets, will be drowned. Our favorite beach will disappear and our prized trout stream will run dry. In short, it means that nature can no longer be protected and preserved, no matter how hard we try.

It means the air, filled with the smoke of a thousand fires, will be grayer, summer days hotter, estuaries slimier. Sunbelters will move back north, Midwesterners will move to Canada, Bangladeshis will have nowhere to go. It means storms and floods will eat the land from under our homes, and some places will cease to exist at all. It means a legacy to our children that will justify contempt for the avaricious generations that left them only the bones of nature to pick at.

A MATTER OF CHOICE

There is another possible outcome, and time still remains to achieve it. Fortunately, signs already abound of a transformation in the way we as individuals and our political leaders look at the world. International relations have entered into a period of extraordinary flux. Nations appear to be moving away from confrontation and toward cooperative engagement over the global threat to the environment. It is, in fact, a prerequisite for addressing the greenhouse problem that this new view of security supplant the fading Cold War perspective. If it doesn't, other issues will appropriate the resources and political attention required to stop the warming.

As in today's world, particular nations will continue to have different roles to play in responding to the global threat; but it is by no means assured that the United States will retain its previous controlling influence if it persists in fourth-wave thinking.

Consider America's technical position. When the Reagan Administration slashed support for renewable energy, some segments of the industry were gutted while others slowed considerably. Orders for rooftop solar hot-water systems dropped 80 percent and the U.S. market share of world photovoltaic sales fell from 80 percent in 1981 to below 50 percent in 1988. Then in 1989 ARCO Solar, with its 15-percent world market share and leadership in nonsilicon technologies, was sold to Siemens A.G. of West Germany. Other nations, perhaps better at reading future economic opportunities, have surged ahead in public spending on photovoltaics: $47 million by West Germany, $54 million by Japan in 1988. If the proposed U.S. 1990 budget is approved, Italy will move ahead of us as well.

As of today, the solution to global warming presents a competitive opportunity that the United States is forgoing. Several Japanese ministries recently organized teams to assess the possibilities, and a group from Mitsubishi roamed worldwide to do the same. When a 20 percent carbon-dioxide reduction was discussed at a 1989 conference, the West German environment minister noted the economic advantages in getting there first. There is plenty of money to be made on the inevitable transition away from fossil fuels. The only question is who will earn it.

Deeper, structural problems will also inhibit the United States in the technology-solution area. We pointed out in chapter 9 that after 1870, in the short space of thirty years, the United States shifted from a wood-based to a coal-based economy. But it is far easier to transform an energy system while a new infrastructure is being built. From 1880 to 1900 alone, the level of industrialization of the United States more than doubled and total fuel use climbed about 80 percent. Similarly, innovations in transportation may be easier to achieve when an economy isn't bound tightly to an existing technology, the way this country is now to the automobile.

This nation's difficulties stem in part from an array of unresolved social problems, which will inhibit our pursuit of greenhouse solutions. Racial issues remain a serious obstacle to improving our system of education and worker training. The perpetuation of an underclass and related problems like drug addiction will continue to drag on our economy and impose constant demands on resources. The political conflict engendered by these and other divisive issues, such as abortion, may fragment the body politic, blocking the pursuit of any cooperative agenda. In other words, current social and political arrangements will make it harder to take advantage of technological opportunities. Until this disjunction is overcome, the United States will remain stuck in the fourth wave, with limited capacity either to implement solutions here or market them abroad.

In addition, innovation is, in some measure, driven by necessity, and the United States has had little need to innovate in greenhouse-specific areas, as compared to other countries. This nation has its own coal supplies so the development of a national energy policy that encourages alternatives lacks a near-term economic and political rationale.

The United States currently falls short in one other respect. Its political system is less vulnerable to the influence of ascendant, if still minority, opinions. It is often said that the vast majority of Americans want strong environmental protection, but few vote for or against candidates based purely on their environmental records. This is not really the fault of the environmental movement, which has, in fact, had some electoral success. Environmentalists battled a popular president to a legislative standstill during the 1980s, protecting most of the gains of the previous decade. But just as with other issues, it is far easier to block losses than to initiate progress, unless your particular concern happens to embody the majoritarian obsession, which has long been military, not ecological, security. For all practical purposes, the environment has remained a secondary issue at the national level, with the brief exception of the early 1970s.

Because the American political system is not parliamentary, it is hostile to third parties, so the segment of the electorate that would vote pro or con on environmental issues ends up frozen out of the political process. It is different in Europe, where the Greens are a minority and yet have attained critical influence in coalition politics. Another problem here is the absence of proportional representation. For lack of such leverage, movements in the United States do not grow easily; their exercise of power must await a groundswell of support, or the accidental absence of forceful opponents. U.S. leadership toward the Montreal Ozone Protocol was greatly facilitated by the lack of strong domestic opposition by CFC producers. No such ease will characterize the limitation of fossil fuels.

Other nations appear better situated to control or benefit from aspects of the greenhouse solution, as judged by current performance and future necessity. As we have seen, photovoltaic technology is well on its way to joining the list of semiconductor advances invented in the United States but exported to the world by Japan. Just as a lack of domestic oil supplies was one of the forces behind its early move toward the fifth wave, Japan may be expected to continue to focus on new energy techniques. Newly industrialized states of the Pacific Rim, such as South Korea, are also in a good position to provide production capacity for the new, silicon-based economy. Germany may market hydrogen cars and other high-value-added renewable energy systems.

India and certain Third World countries may be able to develop indigenous solar hardware industries and thus become sun farmers and hydrogen exporters to northern countries. These nations may be in a far better position than either the United States or Europe* to produce innovations in transportation because they have yet to

*High-speed trains are an innovative approach to intercity travel that are used both in Europe and Japan but are absent in the United States. Although Europe remains less dependent on the automobile than the United States, it has moved more and more toward a car-dependent society.

be frozen into a particular pattern of transport and settlement. It is unlikely, for example, that advances in long-distance mass transportation will arise in countries that are tied to the car. In contrast, much of India remains untouched by either its fine old railway or any modern roads.

The Western European governments have had a checkered record in the past, but they are rapidly becoming precise barometers of environmental sentiment. The various Green parties and a few vanguard politicians, such as Mrs. Brundtland of Norway, have given political voice to these concerns. The Greens may hold the balance of power in the 1990 elections for the West German Bundestag,* and Mrs. Thatcher's sudden interest in both ozone depletion and global warming in 1989 was partly in response to the anticipated erosion in the Conservative Party's position with the voters.

In mid-1989, a controversy over sweeping environmental regulations brought down the government of the Netherlands, which had mishandled the politics of the situation. Environmental concerns have especially strong political ramifications in the countries of Scandinavia; and now even in France, long the *bête noire* of environmentalism, Green strength in local voting has moved President Mitterand at least to gesture in their direction. Electoral power has also produced tangible results, including a shift now under way within the European community toward tough air-pollution controls. Likewise, European cooperation in negotiation of the Montreal Protocol stems originally from pressure brought by German Greens.

Furthermore, environmentalism has provided a focus for public expression of a broad range of the resentments surfacing during Glasnost in the Baltic states and other parts of the Soviet Union, as well as in Poland, Bulgaria, and Hungary, where the future of a massive Danube dam project is in question after heated protests. Both in Eastern Europe and in West Germany, environmentalism seems to be transforming political life as it energizes people across a broad political spectrum. At long last, environmentalism seems on the verge of becoming a new organizing principle.

In *The Rise and Fall of the Great Powers,* Paul M. Kennedy pro-

*As of November, 1989, West German politics were complicated by the opening of the East.

193

posed that power in the world—both economic and political—is in the process of "Balkanization," after a temporary period of bipolar dominance by the United States and the Soviet Union. As the American share of global wealth and international markets declines, so too will its technical leadership, and even its share of greenhouse-gas emissions.

In a Balkanized, highly competitive world, each country will attempt to exploit its selective advantage. Aside from the United States's political stability, which a nation like China lacks, the unique element this country can contribute to a greenhouse solution, provided we act, is an excellent basic research sector, which may supply the sort of experimentation that led to the discovery of high-temperature super-conductivity, or the first silicon solar cell, or the new CFC substitutes. The first step the United States should take as we approach this new world is to protect this existing advantage by increasing support for research and technical education.

Beyond research, building the capacities that allow a transition to the fifth wave will be expedited if U.S. leaders begin to treat global environmental protection as an organizing principle in the same way that they once did anti-communism. This approach could allow for an economic and political mobilization, which would help carry all of us across the bridge into a new era.

Indeed, the public seems prepared for such a shift. The 1988 presidential election saw candidate Bush successfully exploit the environmental issue against Michael Dukakis, neutralizing the Democrats' advantage despite eight years of brutal antienvironmentalism by the Reagan Administration. Had Dukakis appreciated its potential and acted early, he could have retained the advantage on the environment, and it might have proved decisive. However, campaign rhetoric does not amount to a paradigm shift. The final outcomes of the federal clean-air debate, and the battle over the Los Angeles clean-air plan in particular, will provide important signals of how much political will to deal with tough issues really exists.

How can a leader actually establish the hegemony of a new political principle? For that matter, how was the Cold War frame of mind first propagated? Certainly not by mere proclamation, but rather by a series of mutual antagonisms, from the closing days of World War II through the Berlin crisis, the Korean War, the U-2 incident, the Cuban missile crisis, Vietnam, and Afghanistan. Enmity festered, and out of it grew a longstanding cast of characters who guaranteed continuity in the execution of foreign policy and the propagation of political attitudes. This group consisted of people who advised and negotiated, regardless of the party in power; men like Paul Nitze, who was an influential exponent of the hard line as far back as the Truman era, long before he ever negotiated arms issues for Ronald Reagan. Moreover, sustained popular and political consent was assured by a stream of pork-barrel projects, which flowed from the Defense Department's coffers into key congressional districts.

Some of these elements are already in place with regard to the environment, and others could easily be implemented. Americans are beginning to regard environmental threats as they did Communism, as a fundamental and insidious danger to our survival. Every day seems to bring new confirmation of the risk, from the ozone hole to pesticides, to toxic waste sites, to the whole summer of '88, to the *Exxon Valdez* accident and the oil spills that followed it. World leaders, some sensing political gain and others genuinely concerned, are beginning to pound away on their respective drums, reinforcing the conviction that our very existence is at stake.

But two critical elements are missing: the pork is not yet in the barrel, and the so-called policy experts, whose thinking is wedded to fourth-wave relationships, are still in ascendancy, closing their eyes to the possible links among the environment, national security, and economic development. Dimitri Simes of the Carnegie Endowment complained about the attention given the environment and other nonmilitary subjects now part of State Department negotiations with the Russians, saying, "If we are talking about environmental issues, I don't really see what the Soviets can contribute." The Soviet Union, among other things, emits nearly 20 percent of global carbon dioxide.

A similar problem arises with those who dole out economic advice. Accounting for the economic benefits of environmental protection has proven difficult when today's public health, wildlife, or forests have been at stake, as in the case of air pollution. When it comes to reckoning values across generations, there is no accepted basis for comparing the costs of actions with the benefits. Like the ecologists, economists lack a comprehensive framework with which to assess the greenhouse problem. As a consequence, with a very few exceptions, they have not played a constructive role in the greenhouse debate. They have generally taken a "conservative" approach, regarding the potential damages as manageable and the costs of fossil-fuel reductions as an unreasonable burden to the economy. This attitude of discounting future problems pervaded the 1983 National Academy of Sciences report, for example, because two leading economists, Yale's William Nordhaus and Harvard's Thomas Schelling, were dominant figures on the committee that wrote it. (In 1986, Schelling was quoted as saying, "Even if the worst of the predicted climate changes show up . . . carbon dioxide isn't going to be on my list of the half-dozen things we need to worry about").

It should be very clear by now, however, that the resolution of the global environmental issue desperately requires the attention of economists. Chapter 9 underscores the inseparable nature of economic planning and environmental protection, and economists are indispensable as engineers of the process. Although they may not understand everything about future outcomes, no one else does either. But economists at least know something about squeezing innovation out of the system. We can only expect serious attention to global warming when the Council of Economic Advisers deliberates on the high-tech opportunities in the greenhouse technologies; the Office of Management and Budget concedes that certain types of energy-use regulation can bring economic benefit; and the Council on Foreign Relations lobbies for a Secretary of State with environmental expertise. Perhaps it is a salutary signal that environment was the main concern of the 1989 Group of Seven Economic Summit, and that Paul Nitze's son William now serves as Deputy Secretary of State for Environment, Health, and Natural Resources.

Not only do we need a new generation of policy advisers, we also need a new set of institutions in which they can work. Recently, Mrs. Brundtland suggested that NATO redefine its mission in terms of environmental concerns. Others have proposed that the UN Security Council be reorganized and become a mechanism for developing consensus on environmental matters. The UN Environment Programme surprised its considerable number of detractors by successfully nurturing the Montreal Protocol into existence and, with the cooperation of the UN's World Meteorological Organization, moving aggressively to set a framework in place for negotiation on the global climate. It should be recalled that it was these two organizations, not national governments, that got the political ball rolling on the greenhouse problem by organizing the 1985 Villach meeting. The Environment Programme also cosponsored a later meeting at Villach in 1987, which for the first time elaborated a policy agenda for governments. But the organization remains pitifully underfunded.

How else can government stimulate new ways of thinking? An environmental pork barrel—not pork really, but the meat of the solution—will be necessary to counter the economic pinch some industries will inevitably experience with fossil-fuel reductions. The Manhattan Project and the Interstate Highway Program were two of the largest public-works projects in U.S. history, the latter costing $15 billion in 1987 alone. A greenhouse Manhattan Project, but one decentralized through the private sector and the universities instead of centered in government laboratories, should be created to speed innovations in energy technology. It could be supported in part through a slowly expanding policy of renewable-energy procurement. But the pork should be dangled and doled out gingerly and coupled with regulatory measures (per chapter 10) in order to achieve results. A fate like that which befell the synfuels program could kill renewable energy.

If environmental procurement and research are to expand, the military budget will need to contract. One of the original sources of fourth-wave innovation, military spending is now a technological albatross. As mutual force reductions are already under discussion with the Soviet Union, one way to establish the new paradigm would be to tie reduction in military spending to a timetable for increases in support for environmental technology. In fact, environmental spending might provide the means to a military build-down without domestic economic stress. The United States quickly shifted 30 percent of its productive capacity away from the military after World War II. Some dislocation resulted; but in the present case, more time is available for transition.

Again, we find the United States moving in the wrong direction. In 1989, President Bush proposed a renewed commitment to manned space exploration, apparently to resuscitate defense contractors, who are struggling because the defense budget has stopped growing. It would be wiser to retool these industries for the inevitable competition in the renewable energy marketplace. Solar cells have better prospects than either missiles or manned spacecraft.

There are many other needs, such as housing, that cry out for resources. But the global environment has the potential to attract the sort of broad political support needed for significant changes in our fiscal arrangements, and several other pressing infrastructure needs are congruent with the greenhouse agenda. After all, should billions be infused into renovating the highway system without thinking about the future of transportation? Should more housing be built without long-term urban planning in mind?

A few politicians, Senator Timothy Wirth in particular, have explored the idea of assembling a new coalition of interests that might organize such a political transition. Senator Albert Gore and others have spoken out eloquently and, like Wirth, have submitted legislative proposals. What they haven't yet done is to put the global-environment issue at the center of *each* political and budgetary decision, making it the problem that energizes every other priority.

In essence, our argument is that global environmental protection should become the political rationale for an industrial revitalization. America's old industrial policy, centered on military spending, is now failing. It must be replaced with a policy that motivates the public. As Jacques Gansler, a former Pentagon official, says, "National security sells. Industrial policy doesn't." Perhaps the environment will sell too, because it entails the same potent combination of ethical principle and economic self-interest. Environmental stability is the essence of national security. Of course, environmental protection by itself will not solve problems like high interest rates or low rates of saving. But it could provide the political motivation for broad economic restructuring.

THE MAINSTREAM AND THE MARGINALS

In the past, economic and environmental self-interest have vied with each other because they were seen as competing concerns. This separation is no longer tenable. America cannot move to rescue the environment unless it solves its economic problems, and any permanent solution to its economic problems must simultaneously address its ecological concerns.

Even so, there are substantial obstacles. That corporate officers should care about the environment they will leave to their grandchildren as much as they care about the trust funds they set up for them is as difficult a proposition as the notion that Third World countries will suffer if they don't stop selling off their resources. For both groups there are other, overriding considerations, such as immediate profit or simple physical survival.

Development of a modern educational system, for instance, has a payback time to the economy, in the form of skilled workers, of at least twenty years. Increasing U.S. productivity by bringing in the wasted talent of those people now living at the economic edge would have a handsome, but similarly remote payoff. This problem lies at the heart of many parts of the greenhouse issue itself. The only way it will be dealt with is if societies alter their way of accounting for, and discounting, the future, just as they must find ways to alter consumer discounting of energy-efficiency benefits of the lightbulbs they purchase. The threat of global warming may provide the concrete means for creating this essential revolution in attitudes.

There will be no fifth wave for the American economy unless the marginal members of our society are given a stake in its survival. Northern European countries have temporarily dealt with their underclass by means of the welfare state. But even their safety nets are tenuous and may become more so after Europe unifies in 1992, more tenuous yet as the Iron Curtain goes up and emigration westward increases. Radical social critic Ivan Illich has noted that, since the 1970s, even though economies in the developed world have grown, the number of people left out of the general prosperity has risen. The consequences are visible in the number of homeless in the streets of New York or London and the number of unemployed on the Continent. These people represent the flip side of recent prosperity, but it is not likely that such disparities are sustainable. Sooner or later, they divert attention from everything else, and progress is no longer possible.

199

Unfortunately, global warming would worsen the existing inequalities. Dislocation will press down hardest on those who have the fewest resources to adjust, and their suffering will eventually affect everyone. Yesterday's refugees can haunt the rest of the world for centuries, so history belies the "we can handle it" view of global warming. Great Britain's UN Ambassador Sir Crispin Tickell, who persuaded Mrs. Thatcher of the significance of global warming and is awed by the potential proportions of the greenhouse refugee situation, put it this way: "Even if some people and governments wished to seal themselves off from the rest of the world, they could not do so. In no country or city can the rich fortify themselves for long against the poor."

The fifth wave entails the opportunity to deal with environment and equity simultaneously. After all, there's plenty of work for the unskilled in building hydrogen pipelines, or installing solar modules to energize the world. Renewable energy will permit the modernization of Third World economies without accelerating warming, but these countries cannot make the transition on their own. The developed nations can no longer ignore the social inequities that encourage deforestation, and they cannot afford to monopolize those technological advances in energy production that might overcome global warming.

If this situation requires some new flexibility from the political right, the same will be needed from the left. If social inequities and quarterly profits are the source of a considerable amount of our environmental difficulty, it could also be argued that rejecting "Band-Aid" technological solutions and waiting for the Revolution means that a considerable quantity of greenhouse gases will be lost into the atmosphere. We are not arguing that some level of social restructuring is not in order; in fact, we have said quite the opposite. But the limit of time dictates that we begin by working with what is available now, with our institutions and technologies in place and serving as a bridge, based on a guiding principle that is environmental, not political in nature. No one has any idea where slowing global warming might lead us in terms of political or social reorganization, any more than one could have predicted ten years ago that the right-hand headline on the *New York Times* would one day report a popular insurrection in China, while the left-hand one described the Communist party losing elections in Poland.

In fact, it would be shortsighted to reflect on recent victories and conclude that democracy and capitalism are on an indefinite winning streak. The dustbin of history is filled with economic and political systems that failed to address basic human needs. The lack of a vigorous competitor at this time may only highlight the vulnerability of this system to the current challenge, for it remains entirely uncertain whether the democratic–capitalist world can mobilize to beat the greenhouse effect. If it does not, there will be no rival to blame.

Capitalism has particular imperfections that make it vulnerable in this area. Throughout, we have made the somewhat naïve assumption that businesses would respond to profit and regulatory motivation, and we have placed questions of monopoly and its related political power aside. Perhaps this supposition will be justified by the diverse world of the next few decades, where competition will press in from all sides. But perhaps those who dominate our energy supplies will not relinquish their power so easily. Ultimately, the degree of flexibility displayed by corporations will determine how directly governments need to intervene.

One way or another, intervention in the course of technological evolution is inevitable because the public will eventually demand it. Still, it remains unlikely that government will do much, if anything, without forceful prodding from the nongovernmental organizations that represent the public's interest when government doesn't. For example, the Carter Administration's interest in global warming in 1979 originated not from the government's own environmental experts, but from the worries of a few private scientists and the machinations of environmentalist Rafe Pomerance, now at the World Resources Institute and a key figure behind the scenes in Washington.

Environmentalism, like big business, is in the process of globalization, with information and attitudes exchanged freely across borders and across oceans. Describing the recent tactics of the European Greens, Daniel Cohn-Bendit, now a Green, but formerly "Danny the Red" of the 1968 upheavals in Paris, was quoted as saying, "For every environmental problem we can take the country with the most advanced laws and turn those into the norm for the European Community." One factor influencing the Bush Administration to reverse its opposition to a climate convention (discussed in chapter 10) was a call by the British government, which was facing Green pressure at home, for just such an accord. As EPA Administrator William K. Reilly noted a few weeks later, "There is something of a race on between the [1989 Economic] Summit leaders to see who can be the greenest."

Cooperative international action by environmentalists had been growing since the advent of global issues such as tropical-forest protection and the exploitation of Antarctica. But it was the negotiation of the Montreal Protocol that led to the establishment of regular meetings and a long-term, global political strategy. For example, a single scientist in Australia, who was in a position to understand the fast-breaking research findings, helped other environmentalists sway his government's policy on the ozone layer with technical reports gathered from U.S. counterparts. Out of such cooperation grew the Climate Action Network, which—facilitated by telephone, fax, and computer links—ties several dozen independent groups together on a continuing basis.

Another example is a group of U.S. and Soviet scientists who have set up a computer network called "Greenhouse Glasnost" to explore the implications of climatic change. But the network took more than a year to get going because the U.S. government delayed an export license allowing the Soviets to acquire the necessary personal computers. Such misadventures aside, the greenhouse question will be resolved in a worldwide political arena, with the use of new information technologies and strategies for political action. And just as the new international corporations can pick and choose targets for investment and marketing, playing off one country against another to extract the maximum in concessions, so too will the new environmentalists operate on a global scale.

With different nations playing particular roles in the future, this globalization of political lobbying will be especially important. For example, Japan's place at the forward edge of technology and wealth should, in theory, position it to lead politically as well. The technological choices that Japan makes could exert extraordinary leverage on the future course of our energy systems. If Japan backs photovoltaics, the effect could be similar to events at the turn of this century, when the United States backed the production of automobiles.

But Japan has an equivocal record on the environment. Its controls on urban air pollution are more sophisticated than ours, but its exploitation of tropical forests in other Asian countries and its attitude toward whaling and the ivory trade have at times been disturbing. Yet there are signs that Japan is changing. At the 1989 Economic Summit, it announced a $2.25-billion program of environmental assistance to other countries. Nonetheless, for the time being it appears that Europe is the place to look for political leadership on the environment, while Japan is the place to look for technological leadership.* But this division of interests may slow the search for solutions. One immediate task for U.S. and European environmentalists is to identify and assist their Japanese counterparts.

THE LIFE-STYLE AND THE LIMITS

The greenhouse problem is often cast as a question of life-style. Opinions expressed range from grim (or sometimes gleeful) counterculture peroration on the need to jettison consumerist behavior so as to save the planet, to Panglossian assertions that, with a little techno-fix here and there, everyone could go about their business as usual. But the question of whether life-styles must be significantly altered cannot be answered and may, in fact, be irrelevant.

Think of society just before World War II. Compare it to today's. In many respects, they are like different languages with no dictionary to translate between them. Fifty years ago television, digital computers, jet planes, lasers, and biotechnology hadn't touched people's lives yet, or didn't exist at all. The fourth wave had not yet begun, and the fifth wave wasn't even imagined.

---

*Europe, particularly the economic powerhouse of West Germany, will also be an important leader in technology. From a technical, economic, *and* political perspective, a united Germany would be the world's real environmental power.

The world will change in similar ways over the next forty or fifty years, so it makes little sense to try to foresee life-styles. The greenhouse solution is not, therefore, a fixed set of future technologies and behaviors to be achieved, but instead a set of criteria for directing change. Photovoltaics and hydrogen provide credible but not unique possibilities. It remains uncertain whether hydrogen cars will be as cheap as gasoline ones are now (gasoline itself will be much more expensive, if available at all, by that time), so it cannot be foreseen whether transportation by car will be as broadly available in the industrial countries. Indeed, we may develop entirely new forms of transportation which, like the passenger jet in 1939, we can barely begin to imagine. These changes will occur gradually because the efficiency bridge can be initiated with no sudden shift in life-styles. The 1970s oil shock entailed serious social dislocations because it occurred so quickly—there was no time to develop satisfactory alternatives—and it is dislocation rather than gradual, long-term change that needs to be avoided.

Ours remains largely a technocratic view of the problem, and doubters will say that the essential limits of technology are being ignored. Maybe so. The history of the past two hundred years has seen consumerist, capitalist society run up against limit after limit, only to escape time and time again by developing new technology. Growth was always possible again, as Schumpeter noted, as more and more people were brought new levels of wealth. But can this progress really go on forever? Are the environmental constraints of the fourth-wave industrialization only like running out of wood, or is something deeper going on? Will the system continue to be a brilliant escape artist, or has some immutable limit been reached, making this the system's last stand?

There is no doubt that unless population is stabilized, technological change will never catch up with ecological degradation, and the natural environment will continue to shrivel into nothingness. Ever since the Club of Rome's "Limits to Growth" and the 1972 Stockholm Conference on the Human Environment, we have also heard repeatedly that overproduction and luxury consumption in the industrial countries will push the world to the brink of ecological disaster just as surely as Third World population growth. We doubt that the transition to the fifth wave will end the latter argument.

Nevertheless, we remain far from any necessary limits on energy exploitation. The sun is still pumping much more energy to Earth than we can usefully exploit, even with a rapid rise in demand. Long before our hunger for solar energy exceeds its availability, the waste heat inevitably associated with using energy would begin to affect Earth's climate; but that point, too, is far, far off.

Shortages of certain materials could become critical in the future, but the solar-hydrogen economy has advantages over the current one in that respect because most of its raw materials are readily available and recyclable. Silicon is the second most abundant element in the crust of the earth, and the phosphorous and boron used to dope it in making semiconductors are also common. Water is almost everywhere and so, therefore, is hydrogen. Electrodes in fuel cells may be made of platinum, which is relatively rare and would present limits to a full-blown hydrogen economy, but more common substances such as nickel can also be employed. Long-line electrification of the Third World might drain copper reserves available for wire, but decentralized solar energy would help solve this problem. Terminating fossil-fuel use could then free carbon to be used in fabricating advanced composite fiber materials, which have important applications in constructing lightweight, energy-efficient cars and airplanes.

The nearest limit on industrial expansion is that posed by the greenhouse effect itself. If energy exploitation is not shifted to renewable forms, the next fifty years or so will certainly see a collision between the demands of the environment and those of the economy, as billions of dollars are drained away to defend coastlines, move agriculture, and replace lost timber resources. Nature does not readily tolerate the environmental compromise beloved by lawyers or diplomats; but what a habitable, healthy, and sustainable planet needs—what humanity needs—is a meeting ground between technology and Ludditeism. Technology itself is not antinature; but if nature's concerns are not explicitly taken into account, its consequences will be disastrous. In this sense, the fifth wave provides an opportunity rather than a guarantee. As Robert L. Heilbroner noted in a glum commentary on the future in 1972, "nature will provide the checks, if foresight and 'morality' do not."

The only real limit, then, is that of human attitudes. The problem is not that we are lazy or stupid or merely unwilling to alter our behavior. Rather, it is a matter of cynicism, the abandonment of belief in our ability to reconstruct the world. Even so, popular behavior has changed in response to crises in the past, and the new attitudes have even precipitated action by our leaders on occasion. For example, in the mid-1970s the popular boycott of CFC-containing spray cans led directly to government bans. The success of voluntary recycling in the mid-1980s is beginning to have consequences in the political sphere. The change of sexual behavior in response to AIDS, the decline of smoking, and the alterations in diet that have precipitated the rapid decline of heart disease over the past three decades—all of these developments are encouraging testament to the fact that people can change quickly when they have sufficient information and incentive to do so.

Political changes throughout the world and a new emphasis on pragmatism over ideology have opened a window of global opportunity. It is one that all of us must seize. As Thomas Paine wrote of America more than two hundred years ago, in words that are especially applicable today, "We have it in our power to begin the world over again. A situation similar to the present has not appeared since the days of Noah." The time for us to move boldly is now.

CHAPTER 12

The Beginning

IMAGINE . . . THE YEAR IS 2050. THE WORLD ISN'T PERFECT, BUT CONditions have been improving for decades. Where Dickensian scenes of decay once blighted cities of both developed and Third World nations, a modicum of prosperity has returned. Where desperate masses once gutted the resources of the countryside, leaving an impoverished landscape behind as they emigrated to the urban areas, now local industries powered by local energy sustain populations whose numbers are approaching stability. Where smog once clotted the air, now blue sky has returned. Where forests once stood, they still stand. Where polar bears once roamed, they still do. Where the greenhouse effect once ran out of control, warming has slowed to a tolerable rate and will level out just short of three degrees. Humanity has turned the corner, instead of plunging over the cliff.

The change began in 1992 when the nations of the world, under the guidance of the United Nations Environment Programme, signed an agreement to limit greenhouse gases. Cynics scoffed at a "toothless" accord, but they failed to notice the underlying currents. Battered by foreign economic invasion and constrained by a limited budget, the United States cancelled the Super Collider, the Space Station, and Star Wars, and diverted billions into research and education. By the mid-1990s, it had concluded a series of treaties with the Soviet Union which led gradually to a mutual reduction of forces. But the expected blow to the defense contractors never materialized because the government poured some of the billions of dollars saved into procurement of solar-energy sources, and several large arms suppliers quickly converted to renewable energy development. The trend accelerated when the Great Drought of the late '90s struck and industry began to anticipate the total phaseout of fossil fuels. Now rich nations compete to supply smaller countries with solar cells rather than weapons.

By 2000, the United States had recaptured the leadership in photovoltaic sales it had forfeited a decade earlier. This happened not a moment too soon, for in 2007 a team of Japanese engineers made a breakthrough in applying room temperature superconductivity to magnetic levitation and developed a cost-competitive train capable of averaging 360 miles per hour. With their long experience in high-speed rail transport, Japan quickly dominated this market, which has grown by leaps and bounds since the attempts of the 1990s to limit air pollution led to a revival of rail travel and the slow demise of the gasoline car.

After unanticipated support from Angelenos who were grateful for even the slight reduction in traffic brought about by their 1989 clean air plan, and in the face of immense pressure from oil interests, southern California officials had set a firm date of 2010 for the elimination of all vehicles powered by carbon-based fuels. For once, the remnant U.S. auto industry saw the handwriting on the wall and moved quickly toward a program of electric and hydrogen car development in order to satisfy the anticipated national and global market. In 2007, an independent inventor developed an air battery in a garage in Flint, Michigan, which permitted a 185-mile range for electric cars and which could be recharged in one minute by simple replacement of a metal plate. The last gasoline car rolled off the Ford line in 2016.

The future looks even better, but it may not lie entirely with either the old powers or even with Japan and Germany. The economies of the East Bloc improved significantly around the turn of the century. This trend began in Czechoslovakia when a Green-led government invested loans from the West in technologies that improved the energy efficiency of their basic industries. India and other formerly poor nations are also on a roll that began shortly after the climate accords, when the rich nations realized that the fate of the planet was at risk from the desire for economic expansion in the Third World. They provided them with renewable technologies, free at first, then at a discount, to wean them away from fossil fuels. Now the Thar desert of India blooms with photovoltaic models, and that country is awash in foreign exchange from hydrogen sales to Northern Europe. But this story is far from unique. The solution to the so-called "debt crisis" of the late twentieth century came when governments compelled the banks to reinvest funds in development based on local production and renewable energy. With the emphasis on internal development as opposed to capital export, exploitation of forests slowed markedly and new industries sprouted.

Coming with the same suddenness as the end of the Cold War, near universal economic improvement set in once the decisions were made to jump to high-efficiency, renewable technologies instead of defending the dying dependence on fossil fuels. The heavy investment of the 1990s rebounded a thousand times over to our benefit, and the planet was saved to boot. As it turned out, when we stood on that cliff, gazing at the future, we had everything to gain by the choices we made, and only disaster to lose.

NOTES

Prologue

P. 1 That five of the last nine years (through 1988) have been the hottest in the record of global temperature measurement is based on an analysis performed at the Goddard Institute for Space Studies. The Climate Research Unit at the University of East Anglia (Great Britain) finds that six of the last nine years were the warmest. Reasons for the difference between these two reckonings are discussed in chapter 4.

P. 2 Although global mean temperature is expected to change much faster in the future than it did *on average* during the glacial retreat, which occurred between 10,000 and 15,000 years ago, fossil records suggest there were some brief periods of rapid and erratic climatic change during that time (see chapter 2 on the effect of ocean currents). On at least two occasions the deglaciation was interrupted by rapid cooling followed once again by rapid warming. One of these periods is called the Younger Dryas event and it ended abruptly about 10,700 years ago, at which time the warming rate at the high latitudes may have been comparable to that projected by computer simulations for the next century. It is not clear from fossil and ice-core records that these events were global in nature and in any case they did not last for very long. The critical characteristic of the projected greenhouse warming is the indefinitely extended nature of very rapid change. See W. S. Broecker, "Unpleasant Surprises In The Greenhouse?" *Nature* 328 (1987): 123–126; and W. Dansgaard, J. W. C. White, and S. T.

210

Johnsen, "The Abrupt Termination Of The Younger Dryas Climate Event," *Nature* 339 (1989): 532–536.

P. 2 Earth's temperature declined gradually but erratically after the dinosaurs expired about 65 million years ago, possibly due to a decrease in volcanic emissions of carbon dioxide that would have reduced the atmosphere's greenhouse effect over time. The shifting positions of the continents also affected climate. Glaciation of Antarctica began about 38 million years ago. The temperature probably fell to about 5 degrees warmer than today sometime between 4 and 5 million years ago (*Australopithecus afarensis* is dated at 3 to 4 million years ago), a time when Earth was cooling rapidly. About a million years ago, Earth became cold enough that ice sheets periodically invaded the mid-latitude regions. Since then, the Earth has seen successive glacial and interglacial periods, triggered by its shifting position in orbit where its mean temperature alternated between cold and warm, and ice cover was simultaneously extensive or diminished. Earth is currently in an interglacial period. The previous warm period (interglacial maximum) peaked about 120,000 years ago.

These inferences of temperature from paleoclimatic (fossil) data are rather uncertain. But it can be reliably inferred from ice-core data reaching back 160,000 years that Earth was about 3 degrees warmer than today during the last interglacial temperature maximum. We use "200,000 years" here to denote the period that has been no more than 3 degrees warmer because the temperature of the last interglacial maximum may be inferred from the ice-core data, but the temperature of the previous maximum (over 200,000 years ago) cannot be established this way. In fact, Earth may not have been more than 3 degrees warmer than today for more than a million years. For further details see chapter 2. Also see H. Flohn, "Climate Change And An Ice-Free Arctic Ocean," in W. C. Clark, ed., *Carbon Dioxide Review: 1982* (New York: Oxford University Press, 1982); and S. M. Savin, "The History Of The Earth's Surface Temperature During The Past One Hundred Million Years," *Annual Review of Earth and Planetary Science* 5 (1977): 319–355; Donald Johanson and Maitland Edey, *Lucy, The Beginnings of Humankind* (New York: Simon and Schuster, 1981).

Chapter 1: The End

As we note in the prologue, chapter 1 is not a prediction of the future course of events; rather, it contains one scenario among many possible ones depicting the outcome of greenhouse-gas emissions. The sources used to construct this scenario include computer-generated climate simulations, paleoclimatic fossil records, previous human experience, and the various published assessments by climatologists and ecologists listed below. Our approach has been to depend largely on the type of events that, by the weight of these studies, may be deemed likely to occur. The scenario describes events in particular cities, states, and small geographical regions, while certain sources on which it is based—particularly the computer simulations—are not highly reliable on scales smaller than the continental. For example, there remains controversy over the extent and intensity of future drying in the American plains region. In attributing such occurrences to particular places and times, we take a necessary liberty for the sake of illustration. Nevertheless, the *consequences* of such drying, such as the dust storms and forest fires described, have a firm historical and scientific basis. Similarly, some of the social and political outcomes described depend on the sort of conjunction of events that will never be predictable.

The climatic basis for the scenario is provided by a computer simulation reported by J. Hansen, I. Fung, A. Lacis, D. Rind, S. Lebedeff, R. Ruedy, G. Russell, and P. Stone in "Global Climate Changes as Forecast by Goddard Institute for Space Studies Three-Dimensional Model," *Journal of Geophysical Research* 93 (1988): 9341–64. This model predicts temperature changes that are moderately greater than those predicted by some other models, but slightly less than those predicted by others, as discussed in the prologue. One shortcoming in the model is the assumption of continued growth of CFC emissions, which may not occur, given the enforcement and eventual strengthening of the Montreal Protocol on Substances that Deplete the Ozone Layer, signed in 1987. If CFCs are in fact eliminated rapidly but no other measures are taken on emissions, the schedule of warming for the period of the scenario would slow by a decade or so during the next century. The temperatures in particular cities are based on estimates for those or nearby places provided by the Goddard group, and, in view of the caution given above, must be regarded as illustrations, not predictions. The simulations provide the average climate characteristics of a city for a typical summer in the future. We use these projections to estimate the characteristics of a "hot" summer in the future by increasing moderately the predicted average number of days with temperature exceeding a certain value.

Similarly, we describe drought conditions that may not occur every year but will become more frequent at many places in the future. In constructing our scenario, we not only took into account the Goddard simulation but also several published comparisons of the predictions of the various models, including J. F. B. Mitchell, "The 'Greenhouse' Effect and Climate Change," *Reviews of Geophysics* 27 (1989): 115–40; and J. Jaeger, "Developing Policies for Responding to Climatic Change" (Stockholm:

Beijer Institute, April 1988). Several models predict a drying over large areas of the United States during summer months in the future, though differences arise in estimating its intensity and precise location.

Sea-level rise was estimated from the work of J. Oerlemans, "A Projection of Future Sea Level Rise," *Climatic Change* 15 (1989):151–174. Global sea level has risen five inches over the past century, partly as a consequence of the 100-year trend in temperature.

The human response to these changes is perhaps least certain, and various opinions are presented in the references cited below. A sanguine view is held by some, a deeply pessimistic view by others. Consensus exists that rapid climate change would lead society into dangerous territory. The later years of the scenario picture such a world.

P. 8 In large measure, we have based our global-warming scenario on studies contained in the following works: James G. Titus, ed., *Effects of Changes in Stratospheric Ozone and Global Climate*, 4 vols. (Washington, D.C.: U.S. Environmental Protection Agency, August 1986), particularly contributions by S. Manabe and R. T. Wetherald, "Reduction in Summer Soil Wetness Induced by an Increase in Atmospheric Carbon Dioxide"; James G. Titus, "The Causes and Effects of Sea Level Rise"; A. M. Solomon and S. C. West, "Atmospheric Carbon Dioxide Change: Agent of Future Forest Growth or Decline?"; R. L. Peters and J. D. S. Darling, "Potential Effects of Greenhouse Warming on Natural Communities"; L. S. Kalkstein, R. E. Davis, J. A. Skindlov, and K. M. Valimont, "The Impact of Human-Induced Climatic Warming upon Human Mortality: A New York City Case Study"; and M. L. Parry and T. R. Carter, "Effects of Climatic Changes on Agriculture and Forestry: An Overview."

B. Bolin, B. R. Doos, J. Jaeger, and R. A. Warrick, eds., *The Greenhouse Effect, Climatic Change, and Ecosystems* (New York: John Wiley, 1986), particularly R. A. Warrick and R. M. Gifford, with M. L. Parry, "CO_2, Climatic Change, and Agriculture"; and *Preparing for Climate Change* (Rockville, Md.: Climate Institute, April 1988), the proceedings of the First North American Conference on the subject, convened by the Climate Institute in Washington, D.C. on October 27–29, 1987. Among the papers of interest are Stephen P. Leatherman, "Effects of Sea Level Rise on Beaches and Coastal Wetlands"; Robert Peters, "Effects of Global Warming on Biological Diversity"; C. R. Harington, "The Impact of Changing Climate on Some Vertebrates in the Canadian Arctic"; and H. A. Regier, J. A. Holmes, and J. D. Meisner, "Likely Impact of Climate Change on Fisheries and Wetlands, with Emphasis on the Great Lakes"; Joel Smith and Dennis Tirpak, eds., *The Potential Effects of Global Climate Change on the United States*, 2 vols., draft report to Congress (Washington,

D.C.: EPA, October 1988), with eighteen chapters summarizing a host of studies. Regional chapters deal with California, the Great Lakes, the Southeast, and the Great Plains; and national studies consider water resources, agriculture, forests, biological diversity, air quality, human health, urban infrastructure, and electricity demand.

Jaeger, "Developing Policies," is a summary of discussions and recommendations of the workshops held in Villach, Austria, in September and October 1987 and in Bellagio, Italy, in November of that year. M. Oppenheimer, ed., *Greenhouse Gas Emissions: Environmental Consequences and Policy Responses* (*Climatic Change* 15 (1989), contains the updated papers of the Villach 1987 workshop, particularly M. B. Davis, "Lags in Vegetation Response to Climatic Change"; G. M. Woodwell, "The Warming of the Industrialized Middle Latitudes 1985–2050: Causes and Consequences"; M. L. Parry and T. R. Carter, "An Assessment of the Effects of Climatic Change on Agriculture"; J. E. Bardach, "Global Warming and the Coastal Zone (Some Effects on Sites and Activities)"; J. Oerlemans, "A Projection of Future Sea Level Rise"; P. Vellinga and S. Leatherman, "Sea Level Rise, Consequences and Policies"; E. F. Roots, "Climate Change: High Latitude Regions"; and M. Oppenheimer, "Climate Change and Environmental Pollution: Physical and Biological Interactions."

For the climatic excesses of North America, see "The Changing Atmosphere—Challenges and Opportunities," *Bulletin of the American Meteorological Society* 69 (12 [December 1988]): 1434; and Robert Claiborne, *Climate, Man, and History* (New York: W. W. Norton, 1970), pp. 378–93. Dramatic weather anomalies of the 1930s are documented by Harold W. Bernard, Jr., *The Greenhouse Effect* (Cambridge, Mass.: Ballinger, 1980), pp. 59–67. In looking for climatic analogues for the greenhouse effect, Bernard studied the heat waves and drought that created the Dust Bowl, but he also noted that the 1930s produced some extraordinary hurricanes, tornadoes, and cold spells. For instance, on November 30, 1935, the temperature fell below freezing in Langdon, North Dakota, and remained there until February 29, 1936. For forty-one straight days, the temperature never rose above zero, and Langdon's mean winter temperature was 8.4 degrees below zero. February 1936 was the coldest month ever in U.S. history, but only five months later in July, North Dakota cooked, with a record high in the state set at 121 degrees, while 120-degree records were set in South Dakota, Texas, Arkansas, and Oklahoma. Also, on January 15, 1932, two inches of snow fell on Los Angeles, the first time since 1877 that measurable snow had fallen on the city.

While it is not necessarily the case that global warming will entail a more erratic climate, it should be understood that warmer and dryer conditions in some places at certain times may well be accompanied by extremes in the other direction at other times or places. For example, the ocean current which maintains a moderate temperature in northwestern Europe during winter—could shift as the world warms, exposing that region to a colder winter climate even as the greenhouse effect increases. By contrast, it has been argued that a reduction in the pole-to-equator temperature difference, which is expected in the future, could result in a general decrease in the intensity of the sorts of storms that occur at the mid-latitudes in winter.

P. 9 In a personal communication with RHB, retired farmer Bob Birkeland of Roland sees the town becoming a suburb of Ames, 16 miles south, but if Dust Bowl conditions prevail, this is not at all likely.

In November 1933 the first of the Dust Bowl black blizzards began; and "black rain" fell in New York and "brown snow" in Vermont (see Bernard, *Greenhouse Effect*, p. 37). See also Frederick Lewis Allen, *Since Yesterday: The 1930s in America* (New York: Harper & Row, 1986), pp. 196–214, for a vivid picture of what happened "When the Farms Blew Away." Early in 1937, soil particles from a dust storm originating in the Texas–Oklahoma panhandle reached Canada. Analysis revealed that the finely textured soils deposited by the storm on ice and snow in Iowa came from near Dalhart, Texas. The windblown soils contained no sand, as opposed to more than 90 percent sand in the residual drifting dune left behind in Dalhart. See Hugh H. Bennett and W. C. Lowdermilk, "General Aspects of the Soil-Erosion Problem," *Soils and Men, Yearbook of Agriculture 1938* (Washington, D.C.: U.S. Government Printing Office, 1938), pp. 590–91.

The summer of 1988 provided a glimpse of the potential effects of extreme drought and heat, but the 1930s saw a truly remarkable string of hot and dry years. A recent comparison of the summer of 1988 with other summers over the last 100 years in the Midwest shows that while 1988 was among the most severe, a string of years between 1930 and 1936 saw unprecedented conditions. Conditions during 1930, 1933, 1934, and 1936 were particularly poor for agriculture. The drought of four consecutive years that we describe here would be unprecedented but would be only one step beyond the situation of the 1930s. On the comparison of 1988 with the 1930s drought, see K. E. Kunkel and J. R. Angel, "Perspective on the 1988 Midwestern Drought," *EOS* (5 September 1989): 817. For the discussion of the increase in the likelihood of extreme drought as the world warms, see J. Hansen et al., "Regional Greenhouse Climate Effects," in *Coping With Climate Change* (Washington, D.C.: Climate Institute, 1988), pp. 68–81. This article projects that the probability of extreme drought will increase rapidly around the turn of the century if greenhouse emissions continue to grow.

P. 9 In J. P. Myers's "Report to the Board for the Science and Sanctuaries Division, National Audubon Society, March–May 1989," he notes that "Cheyenne Bottoms is dry this spring. Shorebirds using this critical staging site in northbound migration are leaving with hardly enough fat to fuel a few hundred kilometers of flight, not even close to the 2,000 they must fly to reach the Arctic. . . . the Platte was at its lowest flow in memory. Because almost all the nearby Rainwater Basin wetlands were bone dry, the vast majority of millions of waterfowl migrating through the region used the Platte as their primary roost. And while this demonstrates the critical importance of river flows during drought years, it increases the risk to botulism and other diseases facilitated by high bird densities. . . . These provide sobering glimpses of a Greenhouse World."

P. 10 See "California" in Smith and Tirpak, eds., *Potential Effects,* vol. 1. Pages 4-11 to 4-24 focus on the likelihood of change in the seasonality of the runoff.

P. 11 On a visit to Tule Lake in November 1959, RHB saw several million birds in one day. The indirect consequences on wildlife and forests of the northward migration of agriculture have been discussed by D. Dudek in "The Ecology of Agriculture, Environment and Economics," prepared for the Villach 1987 workshop.

P. 12 See Davis, "Lags in Vegetation Response," on forest-migration rates. See also Leslie Roberts, "How Fast Can Trees Migrate?" *Science* 243 (1989): 735–37; and "Is There Life After Climate Change?" *Science* 242 (1988): 1010–12. Discussions at the Villach 1987 workshop were also instructive.

P. 12 Discussions with H. H. Shugart were particularly useful here. For yellow pine, see also Raymond M. Sheffield, Noel D. Cost, William A. Bechtold, and Joe P. McClure, "Pine Growth Reductions in the Southeast," U.S. Department of Agriculture, Forest Service, Southeastern Forest Experiment Station, Resource Bulletin SE-83 (November 1985). The authors note (page 32), "One hypothesis is that atmospheric deposition of pollutants in various forms is adversely affecting the growth of trees across large areas of the Eastern United States. . . . Atmospheric deposition, for example, may predispose forests to the effects of other factors that reduce tree growth." One such factor is climatic change. Southern forests are also discussed by R. A. Sedjo and A. M. Solomon, "Climate and Forests," in *Greenhouse Warming: Abatement and Adaptation,* ed. N. J. Rosenberg, W. E. Easterling III, P. Crosson, and J. Darmstadter (Washington, D.C.: Resources for the Future, 1989). For a detailed discussion of the interaction of air pollution and climate stress on forests, see Oppenheimer, "Climate Change and Environmental Pollution."

P. 12 We have sought to steer a middle course on southeastern agriculture. See "Southeast" in Smith and Tirpak, eds., *Potential Effects,* vol. 1, p. 6-1, which says that 10 to 50 percent of the region's farmland could be withdrawn from cultivation.

P. 13 For the abundance of male alligators and other possible anomalies, see Stephanie Pain, "No Escape from the Global Greenhouse," *New Scientist* (November 12, 1988): 38–43; Roberts, "Is There Life After Climate Change?"; and Andy Dobson, Alison Jolly, and Dan Rubenstein, "The Greenhouse Effect and Biological Diversity," *Trends in Ecology and Evolution* 4 (3 [March 1989]). These articles deal with papers presented at the previously cited World Wildlife Fund's Conference on the Consequences of the Greenhouse Effect for Biological Diversity.

The account of the Tennessee River valley flood is based on Smith and Tirpak, eds., *Potential Effects,* vol. 1, chap. 6.

P. 13 The possibility of intensification of hurricanes has been discussed by K. A. Emmanuel in "The Dependence of Hurricane Intensity on Climate," *Nature* 326 (1987): 483–85.

P. 13 On Bangladesh and Egypt, see J. M. Broadus, J. D. Millman, S. F. Edwards, D. G. Aubrey, and F. Gable, "Rising Sea Level and Damming of Rivers," in Titus, ed., *Effects of Changes.*

P. 14 On the East Coast sea-level rise, see *Responding to Changes in Sea Level: Engineering Implications* (Washington, D.C.: National Academy Press, 1987). See also Joseph Mitchell, *The Bottom of the Harbor* (Boston: Little, Brown, 1959), pp. 37–38: "Occasionally, a bubble upsurges so furiously that it brings a mass of sludge along with it. In midsummer, here and there in the harbor, the rising and breaking of sludge bubbles makes the water spit and seethe." For the interaction of climate and other factors stimulating algal blooms, see Oppenheimer, "Climate Change and Environmental Pollution."

P. 14 On rats' feet, see T. S. Eliot, "The Hollow Men," in *The Complete Poems and Plays* (New York: Harcourt, Brace, 1952), pp. 56–59. The poets say it before anyone else.

P. 15 An acrid haze from forest fires in the Southeast drifted into the Northeast during November 1987. See M. Freitag, "Forest Fire Haze Stays in New York," *New York Times,* November 10, 1987.

For the life history, including the feeding habits, of the Arctic cod (*Boreogadus saida*), see A. H. Leim and W. B. Scott, *Fishes of the Atlantic Coast of Canada* (Ottawa: Fisheries Research Board of Canada, 1966), pp. 189–90. For the ecology of Arctic waters generally, see L. Zenkevitch, *Biology of the Seas of the U.S.S.R.* (London: George Allen & Unwin, 1963).

P. 15 On the consequences of global warming for the Arctic, see Roots,

"Climate Change"; also Barry Lopez, *Arctic Dreams: Imagination and Desire in a Northern Landscape* (New York: Bantam, 1987), p. 103, reports that a polar bear died in Churchill from trying to eat an automobile battery in the dump.

P. 17 The ozone hole over the Antarctic was first reported by J. C. Farman, B. G. Gardiner, and J. D. Shanklin, "Large Losses of Total Ozone in Antarctica Reveal Seasonal ClOx/NOx Interaction," *Nature* 315 (1985): 207–10.

P. 17 The quotation from geochemist Wallace S. Broecker of the Lamont-Doherty Geological Observatory of Columbia University is taken from his written testimony submitted to the Subcommittee on Environmental Protection, Senate Committee on Environment and Public Works, Washington, D.C., January 28, 1987. It has also been reprinted in D. E. Abrahamson, ed., *The Challenge of Global Warming* (Washington, D.C. and Covelo, Calif.: Island Press, 1989).

Chapter 2: Cause and Effect

P. 18 The quotation from Pope Alexander VI is found in Julius E. Olson and Edward Gaylord Bourne, *The Northmen, Columbus, and Cabot, 985–1503,* Original Narratives of Early American History (New York: Charles Scribner's Sons, 1925), p. 73. For Davis's voyage, see Samuel Eliot Morison, *The Great Explorers* (New York and Oxford: Oxford University Press, 1986), pp. 331–36.

P. 19 See H. H. Lamb, *Climate, History and the Modern World* (London and New York: Methuen, 1982), an essential reference work by the founder and first director of the Climatic Research Unit at the University of East Anglia. See also T. M. L. Wigley, M. J. Ingram, and G. Farmer, eds., *Climate and History, Studies in Past Climates and Their Impact on Man* (Cambridge: Cambridge University Press, 1981), a collection of papers on such subjects as "The Economics of Norse Extinction in Greenland," "The Use of Stable-Isotope Data in Climate Reconstruction," and "History and Climate: Some Economic Models."

P. 19 The quotations from Ludlum are found in John Imbrie and Katharine Palmer Imbrie, *Ice Ages* (Short Hills, N.J.: Enslow, 1979), p. 183.

P. 20 Richard G. Lillard, *The Great Forest* (New York: Alfred A. Knopf, 1948), p. 85. On historical U.S. fuel-use statistics, see *Historical Statistics of the United States* (Washington, D.C.: U.S. Dept. of Commerce, 1989). Mark Twain is quoted from *Mark Twain's Speeches* (New York: Harper & Brothers, 1923), p. 145.

P. 20 The Edison quote is from Wilson Clark, *Energy for Survival* (Garden

City, N.Y.: Doubleday, 1975), p. 34, an excellent reference work, almost encyclopedic in scope, and thoroughly documented.

P. 21 On evidence for the increase in carbon-dioxide levels over the past two centuries, see R. H. Gammon, E. T. Sundquist, and P. J. Fraser, "History of Carbon Dioxide in the Atmosphere," in *Atmospheric Carbon Dioxide and the Global Carbon Cycle,* ed. J. R. Trabalka, DOE/ER-0239 (Washington, D.C.: U.S. Department of Energy, 1985).

On twentieth-century fuel-use data in the United States, see *Historic Emissions of Sulfur and Nitrogen Oxides in the United States from 1900 to 1980,* EPA-600/7-85-009a (Research Triangle Park, N.C.: U.S. Environmental Protection Agency, 1985).

P. 21 The Edison quotation is taken from Peter Collier and David Horowitz, *The Fords: An American Epic* (New York: Summit Books, 1987), p. 34. David Halberstam, in *The Reckoning* (New York: William Morrow, 1986), p. 76, has Edison telling Ford, "Young man, that's the thing! You have it—the self-contained unit carrying its own fuel with it! Keep at it!" For the impact of racing on the development of the automobile and for the interconnections between advances in technology and the rise of sport in the United States, see Robert H. Boyle, *Sport, Mirror of American Life* (Boston: Little, Brown, 1963), pp. 25–36.

P. 22 The Ford quotation on reducing the price of the Model T is from Collier and Horowitz, *The Fords,* p. 64.

P. 22 On the number of cars in Kansas in 1923, and U.S. car production in 1927, see Kenneth T. Jackson, *Crabgass Frontier* (New York: Oxford University Press, 1985), p. 162.

P. 22 The Will Rogers quotation is from Collier and Horowitz, *The Fords,* p. 13.

P. 22 On today's fuel-use statistics, see *Policy Options for Stabilizing Global Climate* (Washington, D.C.: U.S. EPA, February 1989 draft report); and the World Resources Institute and the International Institute for Environmental Development, *World Resources 1988–1989* (New York: Basic Books 1988). The balance of energy comes largely from hydro- and nuclear power. The data reflect 1986 values, but do not include noncommercial biomass energy. The U.S. was responsible for about a quarter of the carbon dioxide emitted by fossil-fuel combustion.

P. 23 On the greenhouse effect and its problematic name, see S. H. Schneider and R. Londer, *The Coevolution of Climate and Life* (San Francisco: Sierra Club Books, 1984), chap. 5; and M. C. MacCracken, "Carbon Dioxide and Climate Change: Background and Overview," in *The Potential Effects of Increasing Carbon Dioxide,* J. B. Smith and D. A. Tirpak, eds., DOE/ER-0237 (Washington, D.C.: U.S. Department of Energy, 1985).

P. 23 Carbon dioxide may directly dissolve in seawater, or may be taken in directly by marine plants, such as phtyoplankton, during photosynthesis. The carbon becomes part of the marine food chain and eventually turns to sediment on the ocean floor.

P. 24 On fossil-fuel emission data, see *The Full Range of Responses to Anticipated Climatic Change,* prepared for the United Nations Environment Programme by the Beijer Institute, Stockholm, 1989, chap. 4. Methods for removing CO_2 from fossil fuels are also discussed in this report. At this point, none have proven practical.

Up to 2 percent of carbon-dioxide emissions originate during cement production. D. J. Dudek, "Offsetting New CO_2 Emissions," Environmental Defense Fund, New York, September 1988, estimates that 90 percent removal of CO_2 at power plants would multiply electricity costs 3 to 6 times.

P. 26 On the ice-core record, see C. Lorius, J. Jouzel, C. Ritz, L. Merlivat, N. I. Barkov, Y. S. Korotkevich, and V. M. Kotlyakov, "A 150,000-Year Climatic Record from Antarctic Ice," *Nature* 316 (1985): 591–96; J. M. Barnola, D. Raynaud, Y. S. Korotkevich, and C. Lorius, "Vostok Ice Core Provides 160,000-Year Record of Atmospheric CO_2," *Nature* 329 (1987): 408–14, on which figure 2.1 is based; and C. Genthon, J. M. Barnola, D. Raynaud, C. Lorius, J. Jouzel, N. I. Barkov, Y. S. Korotkevich, and V. M. Kotlyakov, "Vostok Ice Core: Climatic Response to CO_2 and Orbital Forcing Changes over the Last Climatic Cycle," *Nature* 329 (1987): 414–18. Strictly speaking, the ice-core isotope ratios are used to infer Antarctic cloud-level temperature, which can be converted to surface temperature at Antarctica. This covered a range of about 22 degrees for the span of time in Figure 2.1. We have used other data in the above articles to estimate the corresponding global mean temperature change given in the figure caption.

P. 26 Ice-core and fossil studies suggesting a link between carbon-dioxide levels and climate go back to the 1970s, but accurate determination of both carbon dioxide and temperature together were first made by the Russian-French team working with the Vostok core.

Earth's past orbital variations can be calculated with high accuracy. The first mathematical description of how these variations might cause the coming and going of ice ages was proposed by M. Milankovitch, a Yugoslavian, in 1920. The notion that carbon dioxide changes are implicated in bringing on ice ages goes back at least to Svante Arrhenius and T. C. Chamberlain in the late 19th century (see also discussion on p. 34). The two strands were brought together by studies of ice cores published over the last decade which directly linked carbon-dioxide levels with temperature changes and sunlight variations caused by orbital shifts.

P. 28 On ocean currents and climate feedbacks, see W. S. Broecker, "Unpleasant Surprises in the Greenhouse?" *Nature* 328 (1987): 123–26.

P. 28 Climate feedbacks were investigated as far back as the 1950s and the early 1960s by G. N. Plass, F. Moller and others. These studies are summarized by S. Manabe and R. T. Wetherald in "The Effects of Doubling the CO_2 Concentration on the Climate of a General Circulation Model," *Journal of the Atmospheric Sciences,* 32 (1975): 3–15.

P. 30 The literature on air pollution and forest decline is growing at an exponential rate. Still, the number of pollutants involved and the presence of interacting factors like insect infestation and climate variation has prevented assignment of particular levels of responsibility for the damage to acid rain, ozone, or their precursor pollutants individually. Recent summaries of the literature include J. J. MacKenzie and M. T. El-Ashry, *Ill Winds: Airborne Pollution's Toll on Trees and Crops* (Washington, D.C.: World Resources Institute, 1988); and *Interim Assessment,* vol. 4: *Effects of Acid Deposition* (Washington, D.C.: National Acid Precipitation Assessment Program, 1987), chap. 7. The latter work is notorious for its particular bias away from acid deposition as a causal agent, but the data presented therein are useful.

Other readable sources include Janet Raloff, "Where Acids Reign," *Science News,* July 22, 1989, pp. 56–58; Robert A. Mello, *Last Stand of the Red Spruce* (Washington, D.C. and Covelo, Calif.: Island Press, 1987); William H. Smith, *Air Pollution and Forests: Interactions Between Contaminants and Forest Ecosystems* (New York, Heidelberg, and Berlin: Springer-Verlag, 1981); and F. Herbert Bormann, "The New England Landscape: Air Pollution Stress and Energy Policy," in *New England Prospects: Critical Choices in a Time of Change,* ed. Carl Reidel (Hanover, N.H.: University Press of New England, 1982). According to Bormann, a professor in the Yale School of Forestry and Environmental Studies, and a principal investigator at the Hubbard Brook Experimental Forest, "No part of New England is free from some form of air-pollution stress. Some parts are now in Stage I [ecosystem functions little affected], but a considerable area is probably well advanced into Stages IIA [plants may suffer change in reproductive capacity] and IIB [sensitive species decline]. . . . However, time, future emissions, and vulnerability of individual ecosystems are factors that could lead to Stage III responses [in which trees, tall shrubs, short shrubs, and herbs die off in succession and eventually cause ecosystem collapse]." The indicator sites to watch are mountain forests, especially where the red spruce is in decline. Bormann noted that the ultimate Stage III response "may be hastened by wildfire burning over the accumulations of dead wood left after the death of spruce."

P. 31 On the human health effects of ozone, see M. Lippmann, "Health Effects of Ozone: a Critical Review," *Journal of the Air Pollution Control Association* 39 (1989): 672–95.

P. 31 On the various trace gases, see H.-J. Bolle, W. Seiler, and B. Bolin, "Other Greenhouse Gases and Aerosols," in *The Greenhouse Effect, Climatic Change, and Ecosystems,* ed. B. Bolin, B. R. Doos, J. Jaeger, and R. A. Warrick (New York: John Wiley, 1986).

P. 31 Nitrous oxide does not destroy ozone directly, but reacts in the stratosphere to form NO_x, which does.

P. 32 Ralph Cicerone, in a personal communication to RHB.

P. 33 George Woodwell, personal communication with the authors.

Chapter 3: The Learning Curve

P. 34 For John Evelyn, see Peter Brimblecombe's fascinating study, *The Big Smoke: A History of Air Pollution in London Since Medieval Times* (London and New York: Methuen, 1987), pp. 47–52. Brimblecombe is a lecturer in atmospheric chemistry at the University of East Anglia, and we have drawn from his book for our footnote on smog on page 30 of this book.

P. 34 A thumbnail sketch of the early history of greenhouse-effect studies by J. H. Ausubel is presented in the National Research Council's *Changing Climate* (Washington, D.C.: National Academy Press, 1983), annex 2. Svante Arrhenius's original article, which refers back to Fourier, Tyndall, and other early investigators, is "On the Influence of Carbonic Acid in the Air Upon the Temperature of the Ground," *London, Edinburgh, and Dublin Philosophical Magazine and Journal of Science* (1896): 237–76.

The roster of scientists who worked on various aspects of the problem in the 19th century reads like a Who's Who of science, including (in addition to Arrhenius and Fourier) Tyndall, Ångstrom, and Röntgen. On page 237 of the article cited above, Arrhenius notes Fourier's assertion that "the atmosphere acts like the glass of a hothouse." The Fourier article cited is found in "Les Temperatures du Globe Terrestre et des Espaces Planétaires," *Memoires de L'Academe Royal des Sciences de L'Institut de France* 7 (1824), 569–604. Fourier uses the following analogy:

> It is difficult to know how much influence the atmosphere has upon global mean temperature. There is no systematic mathematical theory to guide us on this question. The celebrated traveler M. de Saussure carried out an experiment which seems particularly relevant. This consisted of exposing to sunlight a vessel covered with one or more panes of highly transparent glass, placed some distance apart, one above the other. The vessel was lined with a thick layer of

blackened cork which served to absorb and retain heat. Heated air was contained on all sides, within the container and in the spaces between panes. Thermometers placed inside the container and in the spaces above recorded the degree of heating in each compartment. This apparatus was exposed to the midday sun. Repeated experiments showed the temperatures recorded by the thermometer inside the container rising to (progressively higher values). [Translation provided by R. Barr, G. Treverton, S. Riou, M. L. Barr and M. Barr.]

P. 35 On Lotka's role in the chronology of the greenhouse theory, see Ausubel, *Changing Climate.* Other details on Lotka are found in the article by N. T. Gridgeman in the *Dictionary of Scientific Biography,* ed. C. C. Gillespie (New York: Charles Scribner's Sons, 1973).

P. 35 G. D. Callendar, "The Artificial Production of Carbon Dioxide and Its Influence on Temperature," *Quarterly Journal of the Royal Meteorological Society* 64 (1938): 223–37. He later contributed "Temperature Fluctuations and Trends over the Earth" to the same journal, vol. 87 (1961): 1–12. See also Gordon MacDonald, "The Scientific Basis of the Greenhouse Effect," in *The Challenge of Global Warming* (Washington, D.C. and Covelo, Calif.: Island Press, 1989), a collection of papers edited by Dean Edwin Abrahamson.

Although Callendar's first paper was disregarded, the warming since the late 1800s was apparently at that time already having marked consequences. A pronounced movement of southern animals into the Arctic occurred during the first forty years of this century, followed by a diminution of summer pack ice in the thirties and forties. In 1912, the common cod first appeared in Greenland waters, and by the 1930s it formed a substantial fishery. Southern birds never reported in Greenland before 1920 included the Baltimore oriole, American avocet, Canada warbler, and the ovenbird; while the pectoral sandpiper, grey plover, and other high-Arctic birds that prefer a cold climate declined in numbers. See Rachel Carson, *The Sea Around Us* (New York: Oxford University Press, 1953), pp. 183–87, in which she states, "It is now beyond question that a definite change in the arctic climate set in about 1900, that it became astonishingly marked about 1930, and that it is now spreading into sub-arctic and temperate regions. The frigid top of the world is very clearly warming up." See also Barry Lopez, *Arctic Dreams* (New York: Bantam Books, 1987), p. 144, for the arrival of American robins as far north as Baffin Island in recent years.

G. Evelyn Hutchinson, "The Biochemistry of the Terrestrial Atmosphere," in *The Solar System,* vol. 2: *The Earth as a Planet,* ed. G. P. Kuiper (Chicago: University of Chicago Press, 1954), pp. 371–433.

P. 36 R. Revelle and H. E. Suess, "Carbon Dioxide Exchange Between

Atmosphere and Ocean and the Question of an Increase of Atmospheric CO_2 During the Past Decades," *Tellus* 9 (1957): 18–27. According to Gordon MacDonald of the Mitre Corporation, who spent some time at Scripps during the 1950s, the Revelle–Suess collaboration on the CO_2 question was fortuitous, for neither was studying climate. Suess was interested in the cosmic rays that produce the carbon-14 isotope in the atmosphere. Revelle was expert in marine sediments, which were the presumed graveyard for carbon removed from the air by the ocean. Suess and others had noted a small decline in the carbon-14 content of new tree rings versus ones that were fifty years older, indicating that the carbon dioxide taken in by plants in recent years was deficient in carbon-14 compared to earlier times. Fossil fuels are lacking in carbon-14 because it disintegrates by radioactivity over the eons of burial. The two scientists proposed that fossil-fuel combustion had gradually diluted the carbon-14 that is produced continually by cosmic rays, by adding the dominant carbon-12 to the atmosphere. In other words, emissions had not been removed completely and immediately by the ocean. From this and other data they surmised that carbon-dioxide levels would grow significantly in the future and affect climate.

P. 36 On projected carbon-dioxide growth see I. M. Mintzer, *A Matter of Degrees: The Potential for Controlling the Greenhouse Effect* (Washington, D.C.: World Resources Institute, 1987).

P. 36 The Darling–Waggoner colloquy is to be found in F. Fraser Darling and John P. Milton, eds., *Future Environments of North America* (Garden City, N.Y.: Natural History Press, 1966), p. 101.

Gordon MacDonald pointed out to MO two other events during this period that he considers significant. He served on a 1964 National Academy of Sciences panel on weather and climate modification. Keeling's measurements were already sufficiently compelling that the group underscored the potential climatic significance of the CO_2 buildup. In 1965, a panel of the President's Science Advisory Committee, delivering the first major U.S. government report on the environment, highlighted the greenhouse effect.

P. 38 Our remarks on deforestation are based on a personal communication from George Woodwell to the authors.

The quote is from A. Lacis, J. Hansen, P. Lee, T. Mitchell, and S. Lebedeff, "Greenhouse Effect of Trace Gases, 1970–1980," *Geophysical Research Letters* 8 (1981): 1035–38.

P. 39 The quotation from Woodwell, MacDonald, Revelle, and Keeling is the opening sentence of their typewritten report, "The Carbon Dioxide Problem: Implications for Policy in the Management of Energy and Other Resources," submitted to the Council on Environmental Quality in July 1979. The quotation and the report are also cited in Rafe Pomerance's helpful chronology, "The Dangers from Climate Warming: A Public Awakening," in Abrahamson, ed., *Challenge of Global Warming*, p. 260. Now senior associate for policy affairs with the World Resources Institute in Washington, Pomerance modestly omits that he was responsible for instigating the report. He became involved in the global-warming issue in 1978 when he came across a report on the larger implications of coal burning while doing research on acid rain for Friends of the Earth. Since then, Pomerance has played an important role in stimulating congressional interest in this issue.

P. 000 The Academy report is the National Research Council's *Changing Climate*. The EPA report, by S. Seidel and D. Keyes, is *Can We Delay a Greenhouse Warming?* The notion that the oceans would slow global warming goes back to earlier studies by Manabe, Stephen Schneider, and others. As a consequence of this lag effect, global temperature has not yet climbed very much, even though the current carbon-dioxide level is already higher than that found in ice cores for the two previous warmer periods 6,500 and 120,000 years ago.

P. 40 Conference statement, "An Assessment of the Role of Carbon Dioxide and Other Greenhouse Gases in Climate Variations and Associated Impacts," Villach, Austria, October 9–15, 1985.

The ozone hole was first reported by J. C. Farman, B. G. Gardiner, and J. D. Shanklin, "Large Losses of Total Ozone in Antarctica Reveal Seasonal ClOx/NOx Interaction," *Nature* 315 (1985): 207–10. Other details were provided by Joe Farman to MO in an interview.

P. 42 "Chemical Industry," *Fortune* 16 (December 1937): 83–170.

For Midgley's obituary, see *Chemical & Engineering News*, November 10, 1944, p. 1896.

P. 43 The quotation about the Pentagon is taken from a brochure, "CFC: An Economic Portrait of the CFC-Utilizing Industries in the United States," published by the Alliance for Responsible CFC Policy and given to RHB as part of the testimony presented by Richard Barnett, chairman of the Alliance, at a joint hearing held by the Subcommittees on Hazardous Wastes and Toxic Substances and Environmental Protection, Senate Environment and Public Works Committee, in Washington, D.C., on May 13, 1987.

P. 44 Lovelock's "boobed" quotation is from Lydia Dotto and Harold Schiff, *The Ozone War* (Garden City, N.Y.: Doubleday, 1978), p. 9. Lovelock's book, *The Ages of Gaia* (New York: W. W. Norton, 1988), deals with his research on CFCs, among other matters. In a personal communication with RHB in August 1989, Lovelock recalled that in the late 1960s, when many thought the Earth was headed for long-term cooling, he served as a consultant to Shell on CFCs. In his report to the company, Lovelock said that CFCs were a greenhouse gas and as such they could be "remarkably potent to offset global cooling"!

See Mario J. Molina and F. S. Rowland, "Stratospheric Sink for Chlorofluoromethanes: Chlorine Atom-Catalysed Destruction of Ozone," *Nature* 249 (1974): 810–12. The "end of the world" remark is from Rowland, in a personal communication with RHB.

P. 45 Rowland's observations about the longevity of CFCs are taken from p. 8 of the written testimony he submitted at a joint hearing before the Subcommittees on Hazardous Wastes and Toxic Substances and Environmental Protection, Senate Committee on Environment and Public Works, Washington, D.C., May 12, 1987.

P. 45 Robert Abplanalp's comments are from a personal communication with RHB. Of Swiss ancestry (his name means "from flat mountain"), Abplanalp invented his aerosol valve in 1949 when business was slow at his machine shop on Gun Hill Road in the Bronx. He became an ogre to liberals during Watergate because of his friendship with Richard Nixon. No front runner, he befriended Nixon after he lost the 1960 Presidential race. Information on aerosol bans was provided by A. Miller of the University of Maryland.

P. 46 The Convention is dated March 22, 1985. A history of this period in the CFC negotiations is given by D. D. Doniger, "Politics of the Ozone Layer," *Issues in Science and Technology* 4 (1988): 86–92. Additional details were provided to MO by J. Hoffman of the EPA.

P. 46 On the ozone expedition, see S. L. Roan, *Ozone Crisis* (New York: John Wiley, 1989). On the development of the U.S. position, see Doniger, "Politics," and Roan, *Ozone Crisis*. Other details were provided to MO by J. Hoffman. The EPA nearly dropped CFCs as objects of potential regulation during the reign of Anne Gorsuch Burford as administrator. A series of congressional hearings played an important role in focusing attention on the atmosphere beginning in 1986 (see Pomerance, "A Public Awakening").

P. 47 What Hodel actually said within the circles of the Administration remains unclear. Roan, *Ozone Crisis*, p. 201, says, "Hodel had not specifically called for the use of sunglasses, sunscreens, hats, and staying indoors." Doniger, "Politics," p. 90, cites "press reports" referring to "Hodel's suggestion that the President be presented with an alternative policy of 'personal protection' and 'lifestyle change' in lieu of a meaningful accord. What this meant in plain language was staying indoors or relying on hats, sunglasses, and suntan lotions for protection."

P. 48 The Montreal Protocol on Substances that Deplete the Ozone Layer, September 16, 1987.

P. 48 The efficiency of ozone destruction is much higher in Antarctica than anywhere else because the super-low temperatures allow the condensation of nitrates and water out of the atmosphere and onto polar stratospheric clouds, which resemble cirrus clouds. Under normal circumstances, nitrate would combine with the chlorine atoms that destroy ozone, so the absence of nitrate from the air means more atomic chlorine is available to destroy ozone. In addition, the ice crystals which compose the clouds provide surfaces that are ideal locations for chemical reactions to proceed, which help maintain this high level of chlorine atoms. Furthermore, at high latitudes, the low level of sunlight is inadequate to stimulate the reformation of ozone, and an air circulation called the polar vortex, prevents much mixing with ozone-rich, low-latitude air until November. Then, mixing fills in the hole and carries low-ozone air around the globe.

Watson is quoted by P. Shabecoff in "Arctic Expedition Finds Chemical Threat to Ozone," *New York Times,* February 18, 1989, p. 1.

On the European CFC position, see C. R. Whitney, "12 Europe Nations to Ban Chemicals that Harm Ozone," *New York Times,* March 3, 1989, p. A1.

P. 50 On the Helsinki meeting, see C. R. Whitney, "80 Nations Favor Ban to Help Ozone," *New York Times,* May 3, 1989. The agreement of Third World countries is contingent on technology transfer arrangements.

John Firor's remarks are from a personal communication to MO.

Chapter 4: Judges

P. 51 James Hansen is quoted from the *New York Times,* June 24, 1988, p. A1; MO, testifying on the panel with Hansen that day, took appreciative note of this remark.

Reid Bryson is quoted on p. 7 of Jonathan R. Laing's article, "Climate of Fear," *Barron's* (February 27, 1989). The weekly's front-page headline for that issue was, "The Greenhouse Effect: Mostly Hot Air?" It is worth noting that *Barron's* sister publication, *The Wall Street Journal,* has long ridiculed the acid-rain problem in editorials that would get an "F" in high school chemistry class. Similarly, *Fortune* has made light of acid rain. Such a dismissive or distorted approach to serious environmental problems does a disservice to these publications' readers, if only by spreading misinformation that may stifle industrial innovation in devising technologies that could lead to solutions to these problems, thereby downplaying new profit opportunities.

Some additional insight into the controversy around Hansen's remarks can be gained from R. A. Kerr, "Hansen vs. the World on the Greenhouse Threat," *Science* 244 (1989): 1041–43, and the responses to this article published in the Letters section of *Science* 245 (1989): 451–52. The title of the article is entirely misleading, however, as becomes clear from a reading of the piece. Furthermore, we would disagree with Kerr that Hansen's comment was the critical event in generating attention to the greenhouse effect in 1988. This position ignores the years of scientific consensus-building, noted in chapter 3, which gave the issue legitimacy; the U.S. heat wave and drought of that summer, which provided a vivid example of the potential consequences of warming; the ocean pollution (needles on the beach) and other environmental problems that created a background of concern; and the troubles with the ozone layer. Popular media had already begun to focus on the greenhouse effect: witness a *Discover* article in 1986 and a *Sports Illustrated* article in 1987. It is probably fair to say that the Hansen statement was important, but that with everything else that had happened, the media would have awakened to the issue anyway.

P. 53 Temperature analyses by the Goddard Institute for Space Studies are reported by J. Hansen and S. Lebedeff in "Global Trends of Measured Surface Air Temperature," *Journal of Geophysical Research* 92 (1987): 13,345–72; and "Global Surface Temperatures: Update Through 1987," *Geophysical Research Letters* 15 (1988): 323–26. The University of East Anglia analysis by P. D. Jones, T. M. L. Wigley, and P. B. Wright is found in "Global Temperature Variations Between 1861 and 1984," *Nature* 322 (1986): 430–34. The difference between the two analyses arises mostly from data taken long ago, before 1900.

P. 53 Stephen Schneider is quoted from a *New York Times* article by William K. Stevens, "With Cloudy Crystal Balls, Scientists Race to Assess Global Warming," February 7, 1989, p. C1.

P. 55 The attempted suppression of Hansen's testimony was described in the *Washington Post* by Cass Peterson, "Bush Urged to Shift Stance on Global-Warmth Conference," May 10, 1989, p. A2.

P. 56 For a discussion of atmospheric circulation patterns, climate, and weather, see S. H. Schneider and R. Londer, *The Coevolution of Climate and Life* (San Francisco: Sierra Club Books, 1984).

P. 57 The switch in Academy position is reflected in "Global Environmental Change: Recommendations for President-Elect George Bush," published by the National Academy of Sciences, the National Academy of Engineering, and the Institute of Medicine, 1988. This communication to the president followed a less formal evaluation than the earlier Academy report. The Villach experts are quoted from J. Jaeger, "Developing Policies for Responding to Climatic Change" (Stockholm: Beijer Institute, April 1988).

P. 58 For discussions by a group of ecologists on the issue of generalization, see J. Roughgarden, R. May, and S. A. Levin, eds., *Perspectives in Ecological Theory* (Princeton, N.J.: Princeton University Press, 1989), particularly the P. Karieva article.

P. 58 For a description of the ecological studies at the Hubbard Brook Experimental Forest, see F. H. Bormann and G. E. Likens, *Pattern and Process in a Forested Ecosystem* (New York: Springer-Verlag, 1979). For further discussions of the causes and effects of excess nitrate discharge from forests, see B. Bolin and R. B. Cook, eds., *The Major Biogeochemical Cycles and Their Interactions* (New York: John Wiley, 1983); and D. Fisher, J. Ceraso, T. Mathew, and M. Oppenheimer, *Polluted Coastal Waters: The Role of Acid Rain* (New York: Environmental Defense Fund, 1988).

P. 58 On the methodology of ecology see Roughgarden, May, and Levin, eds., *Perspectives.*

P. 60 Einstein is quoted from "On the Method of Theoretical Physics," originally published in *Hein Weltbild,* ed. C. Seelig (Amsterdam: Querido Verlag, 1934); reprinted in *Ideas and Opinions* (New York: Dell, 1973).

The role played by the Michelson–Morley experiment in Einstein's thinking is discussed in great detail by G. Holton in *The Thematic Origins of Scientific Thought: Kepler to Einstein* (Cambridge, Mass.: Harvard University Press, 1973), chap. 9.

P. 61 The eclipse episode is related by Ilse Rosenthal-Schneider in R. W. Clark, *Einstein: The Life and Times* (London: Hodder and Stoughton, 1973).

P. 62 For a history of the acid-rain issue, see Robert H. Boyle and R. Alexander Boyle, *Acid Rain* (New York: Schocken Books/Nick Lyons Books, 1983); and C. C. Park, *Acid Rain: Rhetoric and Reality* (London: Methuen, 1987). For a discussion of the U.S.–Canada Memorandum of Intent Concerning Transboundary Air Pollution, see Gregory S. Wetstone and Armin Rosencranz, *Acid Rain in Europe and North America: National Responses to an International Problem* (Washington, D.C.: Environmental Law Institute, 1983). In North America, attention has mainly focused on acidified lakes and streams in the Northeastern states and eastern Canada, but the damage, existing or potential, in fisheries alone extends over a much wider area. For instance, a number of coastal streams and rivers from Nova Scotia to Maryland appear to be at particular risk because they naturally lack sufficient acid-neutralizing capacity, in the form, say, of limestone, to buffer incoming acids. And there is strong evidence indicating that episodic acid "pulses" following a rainstorm are responsible for the poor survival of the young of striped bass, American shad, and other fish in vulnerable spawning tributaries of the Chesapeake Bay, which was only twenty years ago the most productive estuary in the world. Over time, poor survival of the young has caused a collapse in the numbers of adult fish returning to spawn: for example, between 1970 and 1980 the Chesapeake catch of river herring (alewives and bluebacks) dropped by 94 percent. See George R. Hendry, guest ed., "Acidification and Anadromous Fish of Atlantic Estuaries," *Water, Air, and Soil Pollution* 36 (1–2 [September 1987]). The entire issue consists of papers presented at a conference on the subject held by the Hudson River Foundation for Science and Environmental Research in October 1985. For later developments on the Chesapeake, see "Acid Deposition in Maryland, Summary of Results Through 1988" (Chesapeake Bay Research and Monitoring Division, Tidewater Administration, Department of Natural Resources, Annapolis, January 1989).

P. 63 The remarks by Jack Calvert and their effect on reporters were relayed to MO by Elizabeth Barrett-Brown of the Natural Resources Defense Council, who attended the Academy news conference. (The "get off the dime" comment is also quoted by Philip Shabecoff in "Acid Rain Panel Urges Curb on Pollutants in East," *New York Times,* June 30, 1983, p. A16.) Barrett-Brown also attended another news conference that immediately followed the Academy's at which environmentalists, including MO, reacted to the report. Given the usual caution of scientists, he and National Clean Air Coalition Chairman Richard Ayres were both taken aback by Barrett-Brown's comment that nothing the environmentalists told the press had quite the impact of Calvert's remarks.

P. 63 In 1986, the so-called Citizens for Sensible Control of Acid Rain spent more than $3 million fighting controls. See Alexandra Allen's op-ed column, "Blow Away the Foul-Air Lobby," *New York Times,* June 11, 1988, p. A31. A staff attorney for the U.S. Public Interest Research Group in Washington, D.C., Allen reported in a press release dated August 28, 1988, that the two top contributors to this industry front group were Southern Company Services, Inc., of Atlanta, with $2.25 million, and American Electric Power Service Corp., in Columbus, Ohio, with $2.13 million. Despite its name, this so-called citizens' organization had yet to receive a single contribution from a private citizen. In the op-ed column, Allen also noted that between 1981 and 1986 some 125 political-action committees associated with electric utility, coal, and other anti–clean air interests donated more than $15 million to candidates for the House and Senate. Of the ten members of the House Energy and Commerce Committee who received more than $50,000 from these political-action committees, only one cosponsored the 1986 acid-rain bill. In a personal communication to RHB, Allen reported that between 1983 and July 1989,

Citizens for Sensible Control of Acid Rain had total receipts of $7,522,633 and expenditures of $6,537,877.

The Environmental Defense Fund study is Fisher, Ceraso, Mathew, and Oppenheimer, *Polluted Coastal Waters*. Eutrophication means the overenrichment of a water body with nutrients such as nitrogen. The excessive growth of algae can lead to the depletion of oxygen and the death of other marine life.

P. 64 For histories of the ozone-depletion issue, see Lydia Dotto and Harold Schiff, *The Ozone War* (Garden City, N.Y.: Doubleday, 1978); and S. L. Roan, *Ozone Crisis* (New York: John Wiley, 1989).

P. 65 The articles referred to are K. E. Trenberth, G. W. Branstator, and P. A. Arkin, "Origins of the North American Drought," *Science* 242 (1988): 1640–45; and K. Hanson, G. A. Maul, and T. R. Karl, "Are Atmospheric 'Greenhouse' Effects Apparent in the Climatic Record of the Contiguous U.S. (1895–1987)?" *Geophysical Research Letters* 16 (1989): 49–52. The related *New York Times* articles are, respectively, W. K. Stevens, "Scientists Link '88 Drought to Natural Cycle in Tropical Pacific," January 3, 1989, p. C1; and P. Shabecoff, "U.S. Data Since 1895 Fail to Show Warming Trend," January 26, 1989, p. 1.

Schneider is quoted from a conversation with MO and from Kerr, "Hansen vs. the World."

P. 65 Woodwell is quoted from a conversation with MO during the spring of 1987.

The early history of interest in ozone depletion is described by Dotto and Schiff, *Ozone War*. Some additional details were provided by S. Wofsy in an April 1989 interview with MO. Johnston published his paper on the SST, "Reduction of Stratospheric Ozone by Nitrogen Oxide Catalysis from Supersonic Transport Exhaust" in *Science* 173 (1971): 517–22. The year before, Paul Crutzen had identified nitrogen oxides from natural sources as important players in stratospheric ozone chemistry. D. R. Bates and P. B. Hays had noted in 1967 that stratospheric nitrogen oxides originate in nitrous oxide arising from bacterial action at the Earth's surface. In 1971, McElroy and J. C. McConnell identified the process converting N_2O to NO_x in the stratosphere. McElroy then sought the specific sources of N_2O on Earth.

P. 66 See Fisher, Ceraso, Mathew, and Oppenheimer, *Polluted Coastal Waters*, for a discussion of the difficulty encountered in generalizing about entire ecosystems from a few measurements. The problems in further generalization to the whole globe are discussed in B. Bolin, B. R. Doos, J. Jaeger, and R. A. Warrick, eds., *The Greenhouse Effect, Climatic Change, and Ecosystems* (New York: John Wiley, 1986), chap. 4. Wofsy's studies are reported in "Tropical Rain Forests and the World Atmosphere," in *Proceedings of the American Association for the Advancement of Science Selected Symposium* 101, ed. Ghillian Prance (Boulder, Colo.: Westview, 1986). Historical details were obtained by MO during an April 1989 interview with Wofsy.

P. 68 For the Kevles quote and other details on the history of physics in the United States, see D. J. Kevles, *The Physicists* (New York: Alfred A. Knopf, 1978), p. 341.

P. 69 See the record of the hearing of the Subcommittees on Hazardous Waste and Toxic Substances and on Environmental Protection, Committee on Environment and Public Works, U.S. Senate, October 27, 1987, at which MO was one of the witnesses. McElroy's own attitude on the importance of global warming appears to have changed over time. In 1986 he twice mentioned in MO's presence that a slightly warmer climate might even be desirable. But by the 1987 hearings he had become quite disturbed over the ozone hole and delivered a loud and clear call at the hearing for the elimination of CFCs. In June 1988 at the Toronto Conference "The Changing Atmosphere," McElroy delivered an eloquent call for the control of the greenhouse gases.

P. 70 On the Superconducting Super Collider controversy, see, for example, B. A. Franklin, "Texas Is Awarded Giant U.S. Project on Smashing Atom," *New York Times,* November 11, 1988, p. A1.

P. 71 The quote on weapons personnel is attributed to C. Anson Franklin, U.S. Department of Energy spokesman, by Fox Butterfield in "Nuclear Arms Industry Eroded as Science Lost Leading Role," *New York Times,* December 26, 1988, p. A1.

Even with an Einstein on hand to prod it, government acts slowly, as it did in response to the famous letter to Roosevelt about the atomic bomb. On this point, see J. Newhouse, *War and Peace in the Nuclear Age* (New York: Alfred A. Knopf, 1989).

P. 71 The news conference description is provided by MO, who participated.

P. 72 For a history of American environmentalism, see Daniel Worster, *Nature's Economy: A History of Ecological Ideas* (Cambridge: Cambridge University Press, 1985), p. 343.

P. 73 Robert Livingston is quoted from a personal communication to RHB.

Chapter 5: Over the Cliff

P. 75 Ted Koppel's remark was made to MO on "Nightline," September 7, 1988. The show was most notable for Koppel's dogged pursuit of the question of whether that summer's heat wave was due to the greenhouse effect. As the argument in chapter 3 shows, it is hard enough to reach a consensus on the cause of the hundred-year trend, much less on the reason for one unusual, but not unprecedented, heat wave. The summer of 1988 was most important as an object lesson on the potential consequences of global warming. But it was just one more grain of sugar on the floor, and whether it would have fallen where it did without the growing level of these gases simply cannot be answered. The most that can be said is that the greenhouse gases tilted the floor so that some grains would slide a bit in the hot direction.

P. 78 Trends in emissions and air quality are discussed in the annual *National Air Quality and Emissions Trends* reports of the U.S. Environmental Protection Agency. Projections of future emissions are found in *Acid Rain and Transported Air Pollutants: Implications for Public Policy,* Congress of the United States, Office of Technology Assessment, June 1984; and *Interim Assessment: the Causes and Effects of Acid Deposition,* National Acid Precipitation Assessment Program, Washington, D.C., 1987. Automobile emissions will increase over time as a result of a variety of factors, including growth in vehicle miles traveled, which reflects both the number of cars on the road and the amount each one is driven, and the degradation of tailpipe control equipment on each car as it ages. See Laurent Hodges, *Environmental Pollution,* 2d. ed. (New York: Holt, Rinehart, and Winston, 1977), pp. 66–71, for a history of federal air-pollution legislation.

Electric utility companies have been extending the lives of their old, uncontrolled power plants because the cost of new ones has risen sharply since the mid-1970s. As a result, emissions reductions from plant retirements that were anticipated in 1970 will not materialize as expected. This issue is discussed in chap. 4 of the Office of Technology Assessment, *Acid Rain.*

P. 79 Morbidity and mortality statistics are drawn from J. D. Butler, *Air Pollution Chemistry* (London: Academic Press, 1979). For Danora and New York City, see Hodges, *Environmental Pollution,* p. 86.

Some pollutants that do not possess this steady-state property, and which also accumulate for long times in the environment (but not the atmosphere), include PCBs and lead. In addition, the biological consequences of exposure to even steady-state levels of air pollution are cumulative in some cases, and not reversible.

P. 80 With regard to the emissions reduction needed to stabilize the carbon-dioxide levels in the atmosphere and stop the carbon-dioxide contribution to warming, only a crude estimate is possible. The best guess is that more than a 50 percent reduction will be required, at which point the rate of draining of carbon dioxide into the ocean will just balance the rate of emission. This issue is discussed by J. Firor, "Public Policy and the Airborne Fraction," *Climatic Change* 12 (1988): 103–5.

P. 80 Estimates of remaining fossil-fuel reserves on Earth are from E. T. Sundquist, "Geological Perspectives on Carbon Dioxide and the Carbon Cycle," in *The Carbon Cycle and Atmospheric CO_2: Natural Variations Archaean to Present,* ed. E. T. Sundquist and W. S. Broecker (Washington, D.C.: American Geophysical Union, 1985); and in J. R. Trabalka, ed., *Atmospheric Carbon Dioxide and the Global Carbon Cycle* (Washington, D.C.: U.S. Department of Energy, 1985), chap. 4. Combustion of recoverable reserves would quadruple atmospheric carbon-dioxide levels if the fraction of emissions remaining airborne each year didn't change. Estimation of the effect on global mean temperature of fully exploiting these reserves can be done only very crudely because the airborne fraction could increase as emissions increase, and the climatic models are unreliable at very high carbon-dioxide levels in any event. The time for exhaustion of reserves is estimated by assuming the continuation of current rates of growth in fossil-fuel use.

P. 81 Discussions of using forests to sequester carbon-dioxide emissions include G. Marland, "The Prospect of Solving the CO_2 Problem Through Global Reforestation," Oak Ridge Associated Universities, Institute for Energy Analysis, Oak Ridge, Tenn., DOE/NBB-0082 (Washington, D.C.: U.S. Department of Energy, February 1988); and D. J. Dudek, *Offsetting New CO_2 Emissions* (New York: Environmental Defense Fund, 1988).

P. 81 Reversal of gaseous releases or their effects by a variety of speculative methods is discussed in W. J. Broad, "Scientists Dream up Bold Remedies for Ailing Atmosphere," *New York Times,* August 16, 1988, p. C1.

P. 82 Sulfate reflection is mentioned by W. S. Broecker in *How to Build a Habitable Planet* (Palisades, N.Y.: Eldigio Press, 1985).

"Lag time" is also referred to as the transient response, which is discussed by S. H. Schneider and R. Londer, *The Coevolution of Climate and Life* (San Francisco: Sierra Club Books, 1984). The size of the lag is rather uncertain because simulations of the ocean are still unsophisticated. The lag depends on a number of factors, including the emissions rate itself, which varies with time. We choose a value of forty years based on the Goddard Institute model noted in the prologue, which would be typical of the next several decades. That is, the full effect of gases emitted by 2020, for example, would not be felt until about 2060.

P. 83 Discussion of historical temperature variations and paleoclimatic data is available in Schneider and Londer, *Coevolution,* chaps. 2 and 3; and B. Bolin, B. R. Doos, J. Jaeger, and R. A. Warrick, eds., *The Greenhouse Effect,*

Climatic Change, and Ecosystems (New York: John Wiley, 1986), chap. 6. An updated analysis of the implications of rapid warming for forests is presented by M. B. Davis, "Lags in Vegetation Response to Climatic Change," in *Climatic Change* 15 (1989): 75–82. A summary of recent thinking on the response of natural systems is presented by L. Roberts in "Is There Life After Climate Change?" *Science* 242 (1988): 1010–12.

P. 85 The history of the (land-based) ice sheets is discussed by Schneider and Londer, *Coevolution*; as well as by H. Flohn, "Climate Change and an Ice-Free Arctic Ocean," in *Carbon Dioxide Review: 1982*, ed. W. C. Clark (New York: Oxford University Press, 1982).

Chapter 6: Heliotrope

P. 87 A recent summary of the various solar-energy technologies has been prepared by the American Solar Energy Society, *Assessment of Solar Energy Technologies*, ed. D. A. Andrejko (Boulder, Colo.: ASES, May 1989). Other eminently readable summaries include D. Deudney and C. Flavin, *Renewable Energy: The Power to Choose* (New York: W. W. Norton, 1983); and C. P. Shea, *Renewable Energy: Today's Contribution, Tomorrow's Promise* (Washington, D.C.: Worldwatch Institute, 1988). Technical reports on particular systems are available from the Electric Power Research Institute, Palo Alto, California. An early view of renewable energy is given by Barry Commoner in "The Solar Transition," 2 parts, *The New Yorker*, April 23, 1979, pp. 53–98 and April 30, 1979, pp. 46–94.

P. 87 See Andrejko, ed., *Assessment*, for description of the Luz project. The impending termination of certain federal tax credits may undermine this and other renewable energy projects.

P. 87 Descriptions and histories of the photovoltaic technology are given by M. Wolf, "Historical Developments in Solar Cells," in *Solar Cells*, ed. C. E. Backus (New York: IEEE Press, 1976); H. M. Hubbard, "Photovoltaics Today and Tomorrow," *Science* 244 (1989): 297–304; and P. D. Moskowitz, P. D. Kalb, J. C. Lee, and V. M. Fthenakis, *An Environmental Source Book on the Photovoltaics Industry* (Upton, N.Y.: Brookhaven National Laboratories, September 1986). The latter work also discusses the environmental impacts of photovoltaic production.

The Bell Labs work on single-crystal silicon cells is described by D. M. Chapin, C. S. Fuller, and G. L. Pearson, "A New Silicon p-n Junction Photocell for Converting Solar Radiation into Electrical Power," *Journal of Applied Physics* 25 (1954): 676. Wolf notes that using selenium cells, 1 percent was the highest efficiency obtained until 1954. Chapin, Fuller, and

Pearson give one-half percent as the highest efficiency in commercially available cells up to that time.

P. 88 Historical photovoltaic costs and production volume are discussed in Hubbard, "Photovoltaics"; Shea, *Renewable Energy;* and Moskowitz, Kalb, Lee, and Fthenakis, *Environmental Source Book.* The technology is developing rapidly, and current costs are subject to quick change. An excellent summary of recent developments is found in S. Ashley, "New Life for Solar?" *Popular Science* (May 1989): 117–59, in addition to Hubbard, "Photovoltaics." Several periodicals, including *Photovoltaic News,* published in Arlington, Virginia, also track developments. Moskowitz, Kalb, Lee, and Fthenakis, *Environmental Source Book,* contains a list of manufacturers along with details about their particular processes. P. Lehman of Humboldt State University in Arcata, California, provided MO with critical details on new developments.

P. 89 Robert Williams and colleague Joan Ogden have analyzed the potential for solar energy in "Hydrogen and the Revolution in Amorphous Silicon Solar Cell Technology," Center for Energy and Environmental Studies, Princeton University, February 1989.

In addition to the preceding references, the limitations of solar energy are noted by a study group administered by Resources for the Future in *Energy: The Next Twenty Years,* sponsored by the Ford Foundation, chap. 13 [1979]. Although the prices are completely out of date, the general discussion is worthwhile.

P. 90 The land requirements for photovoltaics have been calculated many times, and the values change as photovoltaic efficiency improves. The value cited here is similar to one given in Ogden and Williams, "Hydrogen," who point out that world fossil-fuel requirements could be satisfied by utilizing photovoltaics covering less than 2 percent of the Earth's desert area. We assume horizontal collectors; they assume tilted ones and a slightly higher conversion efficiency.

P. 91 The photovoltaic-hydrogen concept and the uses of hydrogen are described in great detail by J. O'M. Bockris, *Energy Options: Real Economics and the Solar-Hydrogen System* (Sydney: Australia and New Zealand Book Co., 1980), but the prices given are entirely out of date. Recent discussions of the photovoltaic-hydrogen system include Ogden and Williams, "Hydrogen," and C.-J. Winter and J. Nitsch, eds., *Hydrogen as an Energy Carrier* (Berlin: Springer-Verlag, 1988). Several conversations with Bockris were extremely useful in writing this section.

The history and current status of fuel cells is discussed in A. J. Appleby, ed., *Fuel Cells: Trends in Research and Applications* (Washington, D.C.: Hemisphere, 1987). See also Winter and Nitsch, eds., *Hydrogen,* and the summary given by J. Holusha, *New York Times,* July 26, 1989, p. D1. Information on Alstrom is from an interview with J. O'M. Bockris.

P. 91 Hydrogen safety is discussed in the preceding references and in J. Hord, "Is Hydrogen Safe?" NTIS PB-262 551 (Boulder, Colo.: National Bureau of Standards, October 1976).

P. 92 Automobile emissions from hydrogen fuel are discussed in *Alternative Drive Concepts* (Munich: BMW AG Press, December 1987), which reports on both BMW's hydrogen and its electric prototype vehicles.

P. 92 See Bockris, *Energy Options,* or Winter and Nitsch, eds., *Hydrogen,* for discussions of hydrogen manufacture.

P. 93 M. Miller, "Not So Bad After All? Nuclear Power Revisited," *Newsweek,* July 25, 1988, p. 65.

P. 94 The costs of Nine Mile Point Unit 2 were obtained from C. Guinn of the New York State Energy Office; they were converted to kilowatt-hour costs by C. Komanoff in a personal communication with MO.

For costs of small-scale options like low-head hydro dams and cogeneration, see, for example, *New York State Energy Plan,* New York State Energy Office, Department of Public Service, Department of Environmental Conservation, Albany, May 1989, draft.

P. 94 Energy-efficiency improvements in a U.S. context are discussed by R. C. Cavanagh, "Least-Cost Planning Imperatives for Electric Utilities and Their Regulators," *Harvard Environmental Law Review* 10 (1986): 299–344; A. H. Rosenfeld and D. Hafemeister, "Energy Efficient Buildings," *Scientific American* 258 (1988): 78–85; and C. Flavin and A. B. Durning, *Building on Success: The Age of Energy Efficiency* (Washington, D.C.: Worldwatch Institute, March 1988). A global view is taken by J. Goldemberg, T. B. Johansson, A. K. N. Reddy, and R. H. Williams, "An End-Use Oriented Global Energy Strategy," *Annual Review of Energy* 10 (1985): 613–88. The arguments in this article are summarized in two reports by the same authors, "Energy for Development" and "Energy for a Sustainable World," published in September 1987 by the World Resources Institute, Washington, D.C.

P. 95 The California case history is examined by D. Roe in *Dynamos and Virgins* (New York: Random House, 1984).

P. 96 See Cavanagh, "Least-Cost Planning," for a discussion of energy-efficiency efforts in the Pacific Northwest.

P. 96 A comparison of nuclear power versus efficiency improvements is presented by B. Keepin and G. Kats in "Greenhouse Warming: Comparative Analysis of Nuclear and Efficiency Abatement Strategies," *Energy Policy* 16 (1988): 538–60.

Costs of coal-fired generation were obtained from C. Komanoff.

Wolf cites $400 per watt as the cost of electricity from early solar cells. Using one kilowatt as the typical peak electricity demand for a house without electric heating, we find a cost of $400,000.

The cost escalation of nuclear-power plants is discussed in C. Komanoff, "Assessing the High Costs of Nuclear Power Plants," *Public Utilities Fortnightly,* Arlington, Va. (Oct. 11, 1984). The Chronar Corporation's photovoltaic-electricity cost projection is based on a press release of September 8, 1988, Princeton, N.J., and subsequent discussions between MO and Chronar's staff. The Luz costs are discussed in Andrejko, ed., *Assessment.*
P. 97 Efficiency of photovoltaic cells is constantly improving. In addition to Hubbard, "Photovoltaics"; Shea, *Renewable Energy;* Moskowitz, Kalb, Lee, and Fthenakis, *Environmental Source Book;* and Ashley, "New Life"; see also R. Pool, "Solar Cells Turn 30," *Science* 241 (1988): 900–901.

Ogden and Williams, "Hydrogen," contains an extended series of cost estimates for future photovoltaic-hydrogen systems.
P. 98 Komanoff, "Assessing the High Costs," estimates that reactor cancellations and unexpected cost overruns in producing nuclear power each cost ratepayers about $50 billion (in 1984 dollars) by the mid-1980s.
P. 98 The security problems surrounding a "nuclearized" world are discussed by R. H. Williams and H. A. Fiveson in "Diversion-Resistance Criteria for Future Nuclear Power," Energy and Environmental Studies Institute, Princeton University, May 1989.
P. 99 See Moskowitz, Kalb, Lee, and Fthenakis, *Environmental Source Book,* for discussion of photovoltaic-production safety issues.
P. 99 For hydrogen projects now under way see P. D. Hoffmann, D. Hunter, and S. Ushio, "Hydrogen Research: Back in the Spotlight," *Chemical Engineering* (October 26, 1987): 26–28.
P. 100 Gas-cooled reactors are discussed by H. M. Agnew, in "Gas-Cooled Nuclear Power Reactors," *Scientific American* 244 (1981): 55–63. They were discussed in an upbeat article by W. J. Broad in the *New York Times,* November 15, 1988, p. C1; M. L. Wald reported the impending shutdown of the 330-mw reactor at Fort Vrain, Colo., in the *New York Times* on December 8, 1988, p. B18. On May 11, 1989, S. Dickman reported in *Nature* 339 (p. 90) on the West German reactor.

The raid on the Federal treasury began with the September 1989 announcement by the Department of Energy of a $50-million contract with Westinghouse to develop a cleaner and safer reactor.

For a discussion of changing attitudes about nuclear power in Europe and the Soviet Union, see "State of the World, 1989," Worldwatch Institute, Washington, D.C., pp. 5–6, and related references. This section is also based on the personal observations of MO during several trips to Europe, including the Soviet Union, since the Chernobyl incident.

P. 100 For a discussion of nuclear-power programs abroad, see Keepin and Kats, "Greenhouse Warming."

P. 102 For a discussion of extraction of biomass energy using modern technology, see Shea, *Renewable Energy;* and Deudney and Flavin, *Renewable Energy.* For the global perspective on biomass, see Goldemberg, Johansson, Reddy, and Williams, "End-Use Oriented," "Energy for Development," and "Energy for a Sustainable World."

P. 103 For the Brazilian ethanol program, see D. L. Bleviss, *The New Oil Crisis and Fuel Economy Technologies* (New York: Quorum Books, 1988), in addition to the references above on biomass.

P. 103 If fossil fuel, rather than renewable energy, is used to power the process, including cultivation, which produces the ethanol, then some carbon dioxide will, of course, be emitted. For example, Brazil's ethanol program is powered by fossil fuels, which consume the equivalent of 18 percent of the energy content of the ethanol produced (private communication with J. Goldemberg).

Ogden and Williams, "Hydrogen," discusses the land- and water-use requirements of biomass energy production.

P. 104 There have been many estimates of the time needed to put new energy technology in place, but one was recently provided specifically in the greenhouse context by the British government's Energy Technology Support Unit in a paper entitled "Abatement of Greenhouse Gases in the U.K.," prepared for the April 26, 1989 briefing of Prime Minister Margaret Thatcher (see chapter 10). Dr. Ken Currie delivered the estimate that in a modern economy, it would require thirty years to cut emissions of carbon dioxide in half. Winter and Nitsch, *Hydrogen,* provides a somewhat longer timeline for a hydrogen system. We discuss this issue further in chapter 9.

P. 104 The Toyota AXV is discussed by Bleviss, *New Oil Crisis.* Fuel economy will be discussed in more detail in chapter 7.

P. 105 The SunFrost specifications were obtained from P. Lehman of Humboldt State University. The company's model is comparable to a standard commercial refrigerator, but is not fully automatic in self-defrosting.

Bleviss, *New Oil Crisis,* discusses the relation of fuel efficiency to air emissions.

For a discussion of the development of the incandescent lightbulb, see T. K. Derry and T. I. Williams, *A Short History of Technology* (Oxford: Oxford University Press, 1960), p. 33. This book is the "sequel" to a much larger work, Charles Singer et al. *A History of Technology,* 5 vols. (Oxford: Oxford University Press, 1954–58).

P. 106 On fluorescent lamps, see M. A. Cayless and A. M. Marsden,

Lamps and Lighting, 3d ed. (London: Edward Arnold, 1983). A typical fluorescent-tube lamp operates as follows. An electrical discharge excites atoms of mercury vapor within the lamp tube, causing them to emit invisible ultraviolet radiation. These rays are absorbed by a phosphorescent coating on the inside surface of the tube, which converts them to visible light. The entire process converts electrical energy to light with four to six times the efficiency of a standard incandescent lamp.

P. 106 A. H. Rosenfeld of the Lawrence Berkeley Laboratory has analyzed the savings from compact fluorescent lightbulbs, which would vary depending on local electricity rates. A technical description of these bulbs is given by E. Rasch, "Properties and Performance of Compact Fluorescent Lamps with Integral HF Electronic Ballasts," *Lighting Design and Application* (September 1987): 28–34. The description of the Sylvania Capsylite is courtesy of P. Miller of the Environmental Defense Fund and Edison Price.

P. 107 The global energy-savings potential of efficiency and its greenhouse-gas implications are discussed by Keepin and Kats, "Greenhouse Warming"; and in the National Energy Efficiency Platform of the American Council for an Energy Efficient Economy, Washington, D.C., June 1989.

Amory B. Lovins provided an early discussion of energy efficiency in "Energy Strategy: The Road Not Taken?" *Foreign Affairs* 55 (October 1976): 61–96. This has been updated in discussions with MO. The Monsanto executive cited was Senior Vice-President Harold Corbett, and the comment was made to MO.

P. 107 For U.S. energy use over the past fifteen years, see W. U. Chandler, H. S. Geller, and M. R. Ledbetter, *Energy Efficiency: A New Agenda* (Washington, D.C.: American Council for an Energy Efficient Economy, 1988). For California data see World Resources Institute and the International Institute for Environment and Development, *World Resources, 1988–89* (New York: Basic Books, 1988).

P. 108 The Southern California Edison program is discussed in Flavin and Durning, *Building on Success.*

P. 108 See Chandler, Geller, and Ledbetter, *Energy Efficiency,* for comparative energy efficiency of different nations; and Goldemberg, Johansson, Reddy, and Williams, "End-Use Oriented," "Energy for Development," and "Energy for a Sustainable World," for information on appliances in particular nations.

P. 108 About eighty percent of total global energy use, and eighty-eight percent of commercial energy use, which excludes much of rural biomass energy, comes from fossil fuels. The estimate that global energy use can remain constant between now and 2025 comes from Gordon Goodman, director of the Stockholm Environmental Institute, based on analysis performed for the World Commission on Environment and Development.

Ford's comment on time wasted is found in Peter Collier and David Horowitz, *The Fords: An American Epic* (New York): Summit Books, 1987) p. 64.

See Chandler, Geller, and Ledbetter, *Energy Efficiency,* for U.S. investment in the electric utility sector. This issue is also discussed by R. B. Zevin, *A Greater Good: Potentials for an Intelligent Economy* (Boston: Houghton Mifflin, 1983).

P. 109 The fraction of the Gross National Product devoted to energy is calculated by Chandler, Geller, and Ledbetter, *Energy Efficiency.*

Chapter 7: Sic Transit

P. 111 The quote on the "American engine" is from J. J. Flink, *America Adopts the Automobile, 1895–1910* (Cambridge, Mass.: MIT Press, 1970), p. 281. Quotes on cylinder capacity and horsepower are also provided by Flink. The original citations are *Scientific American* 95 (1906): 442; and *World's Work* 5 (1903): 3305, respectively.

P. 112 The Model N's fuel economy is from Flink, *America Adopts.* Recent fuel-economy data are from C. J. Calwell, "The Near Term Potential for Simultaneous Improvements in the Fuel Efficiency and Emissions of U.S. Automobiles" (M.A. thesis, University of California, Berkeley, May 1989).

The dimensions of the beast of a 1974 Buick Le Sabre come from Joe Olszewski, General Sales Manager, McCallum Chevrolet–Pontiac–Buick Inc., Fishkill, N.Y., in a personal communication to RHB.

On Third World automobile growth, see "State of the World, 1989," Worldwatch Institute, Washington, D.C., chap. 6.

P. 112 R. Register, *Ecocity Berkeley: Building Cities for a Healthy Future* (Berkeley, Calif.: North Atlantic Books, 1987), p. 8.

On the socioeconomic web, see K. T. Jackson, *Crabgrass Frontier: The Suburbanization of the United States* (New York: Oxford University Press, 1985), a superb synthesis of social history, economics, transportation, and architecture, essential to any understanding of contemporary American society. See also J. J. Flink, *The Car Culture* (Cambridge, Mass.: MIT Press, 1975); and idem, *The Automobile Age* (Cambridge, Mass.: MIT Press, 1988).

P. 113 The number of cars is taken from Flink, *America Adopts.*

For Cugnot's steam-carriage, see T. K. Derry and Trevor I. Williams, *A Short History of Technology* (Oxford: Oxford University Press, 1960), pp. 331–32.

See Flink, *Car Culture*, on the history of early steam-powered road vehicles. Railroads were a superior technology at the time, but their complete dominance over road vehicles was assured in the United States by government subsidy of track expansion and in Britain by the speed and flag requirements of the Locomotives on Highways Act of 1865, which was pressed by railway interests and was finally repealed in 1896.

On Blanchard's vehicle, see R. E. Anderson, *The Story of the American Automobile: Highlights and Sidelights* (Washington, D.C.: Public Affairs Press, 1960).

P. 114 The speed record is found in R. H. Boyle, *Sport: Mirror of American Life* (Boston: Little, Brown, 1963).

On the steam and electric cars and their limitations, see A. Jamison, *The Steam Powered Automobile: An Answer to Air Pollution* (Bloomington: Indiana University Press, 1970); and D. J. Santini, "Commercialization of Major Energy-Enhancing Vehicular Engine Innovations: Past, Present, and Future Micro- and Macroeconomic Considerations," in *Energy Analysis and Alternative Fuels and Vehicles,* Transportation Research Record 1049 (Washington, D.C.: Transportation Research Board, National Research Council, 1985); in addition to Flink, *America Adopts* and *Car Culture*.

On Kettering and the automatic ignition, see Derry and Williams, *A Short History,* p. 719. On its relation to the electric car, see Santini, "Commercialization."

P. 114 On development of the internal-combustion engine and its use in cars, see Flink, *America Adopts* and *Car Culture*. Strictly speaking, Lenoir developed the first commercially successful internal-combustion engine. Others had patented different versions earlier. See also Derry and Williams, *Short History,* which notes on p. 605 that Gottlieb Daimler's first petrol engine, "patented in 1885, was a single-cylinder vertical machine, air-cooled, working on the Otto cycle. . . . After 1893 Daimler, and subsequently other manufacturers, used the modern float-feed carburetor invented by Wilhelm Maybach." The original Lenoir and Otto engines ran on gas, not liquid fuel. From about 1885 onward, the Otto engine began to compete with the steam engine in factories for certain uses. At about the same time, liquid fuel was replacing gas. There are minor discrepancies in dates among the various sources.

P. 115 For the effect of Charles Duryea's victory on Ford, see Boyle, *Sport,* p. 33.

P. 115 On the "1,000 small shops," see Flink, *Car Culture,* p. 16. The bicycle, which became widespread after 1885, and its proponents played an important role in the development of the automobile, both as a source of technological and manufacturing ingenuity and by literally paving the way for improved roads. On the bicycle and the automobile, see Flink, *Automobile Age*.

P. 116 The Chicago health commissioner is quoted from Flink, *America Adopts*, p. 105, who cites "Merely a Matter of Time," *Motor Age* 5 (December 1901): 5.

See Jackson, *Crabgrass Frontier*, on the set of decisions favoring automobiles over mass transit; also R. A. Caro, *The Power Broker: Robert Moses and the Fall of New York* (New York: Vintage Books, 1975).

P. 117 Jackson, *Crabgrass Frontier*, p. 170. On the decline of ridership, see also G. W. Hinton, "The Rise and Fall of Monopolized Transit," in *Urban Transit: The Private Challenge to Public Transportation*, ed. L. A. Lane (Cambridge, Mass.: Ballinger, 1984).

See Caro, *Power Broker*, chap. 24, for Moses' attitudes regarding mass transit.

P. 118 The quote is from L. J. White, who provides an incisive examination of the American automobile industry on the brink of decline in *The Automotive Industry Since 1945* (Cambridge, Mass.: Harvard University Press, 1971), p. 274.

P. 119 The quote is from A. W. Bruce, *The Steam Locomotive in America* (New York: W. W. Norton, 1952), p. 391; on the technical development of the steam locomotive, see also J. H. White, Jr., *A History of the American Locomotive/Its Development: 1830–1880* (New York: Dover, 1979). On the transition to electric power, see Santini, "Commercialization," as well as Bruce, *Steam Locomotive*.

P. 120 On the speed of shifts in technology, see J. O'M. Bockris, *Energy Options: Real Economics and the Solar-Hydrogen System* (Sydney: Australia and New Zealand Book Co., 1980), chap. 3.

Our comments on the maturation of automotive technology arise from an MO interview with J. J. Flink. The motor reached its current form by the late 1920s, though the development of automatic transmissions and hydraulic brakes, along with improvements in carburetion and suspension, occurred during the 1930s. The postwar period saw incorporation of these technologies into road models, but new advances were generally restricted to production methods, not automotive engineering. In the mid-1970s, the American industry was finally jolted into making technical change by environmental restrictions, energy shortages, and foreign competition. See also L. J. White, *Automotive Industry*.

On compounding in the automobile, see D. L. Bleviss, *The New Oil Crisis and Fuel Economy Technologies: Preparing the Light Transportation Industry for the 1990's* (New York: Quorum Books, 1988); and Santini, "Commercialization."

P. 122 The Perris Smogless Automobile Association is cited by Bockris, *Energy Options*, p. 275.

P. 122 For technical descriptions of the hydrogen car and its limitations, see Joan Ogden and Robert Williams, "Hydrogen and the Revolution in Amorphous Silicon Solar Cell Technology," Center for Energy and Environmental Studies, Princeton University, February 1989; Bockris, *Energy Options;* M. A. DeLuchi, "Hydrogen Vehicles," in *Alternative Transportation Fuels: An Environmental and Energy Solution* (New York: Quorum Books, 1989); and *Alternative Drive Concepts* (Munich: BMW AG Press, December 1987). These were supplemented by MO's interviews with Ogden and Williams.

P. 123 The Bavarian plant is described by P. Hoffmann, D. Hunter, and S. Ushio in "Hydrogen Research: Back in the Spotlight," *Chemical Engineering,* October 26, 1987, pp. 26–28. Fueling the TU-155 with hydrogen is described by P. Hoffmann in "The Fuel of the Future Is Making a Comeback," *Business Week,* November 28, 1988, pp. 130–31. The West German hydrogen network is described in C.-J. Winter and J. Nitsch, eds., *Hydrogen as an Energy Carrier* (Berlin: Springer-Verlag, 1988), p. 253.

On electric vehicles, see M. A. DeLuchi, Q. Wang, and D. Sperling, "Electric Vehicles: Performance, Life-cycle Costs, Emissions, and Recharging Requirements," *Transportation Research* 22A (forthcoming); *Alternative Drive Concepts;* Bleviss, *New Oil Crisis;* and W. Hamilton, *Electric Automobiles: Energy, Environmental, and Economic Prospects for the Future* (New York: McGraw-Hill, 1980).

P. 124 On fuel-economy regulations, see Calwell, *Near Term Potential;* and J. H. Cushman, "Tougher Fuel Economy Rules Planned, in Shift from Reagan," *New York Times,* April 15, 1989, p. 1.

P. 125 On the Ford Escort, see E. J. Horton and W. D. Compton, "Technological Trends in Automobiles," *Science* 225 (1984): 587–93. This article, written by two Ford executives, provides a telling glimpse of the fuel-economy status of automobiles and their future potential. A mid-1980s model Escort expelled 83 percent of the energy of combustion as heat from the tailpipe and the radiator in approximately equal amounts. Only 13 percent of the combustion energy was used for propulsion, 4 percent for minor purposes like powering accessories. In contrast, at the end of the article, the authors describe the car of the future as follows: "So what might an 'average' vehicle of the late 1990's be like? It could be a four- or five-passenger vehicle in the 2000-pound inertia weight class with an aerodynamic drag coefficient of 0.20 or less. Its fiber and plastic composite body panels (assembled by adhesive bonding) would ride on an electronically controlled suspension system, with the driver selecting either a boulevard ride or a stiffer ride more appropriate for freeway cruising. Electronics would control a turbocharged, ceramic, adiabatic diesel engine and continuously variable transmission to provide smooth effortless performance and fuel economy in excess of 100 miles per gallon on the highway. The increased costs of such a combination must be weighed against the savings in fuel cost. The appearance of such a vehicle will depend strongly upon the price of fuel."

P. 125 On specific fuel-economy measures, see Bleviss, *New Oil Crisis,* in addition to Horton and Compton, "Technological Trends." For a discussion of why they won't happen easily, see F. Von Hippel and B. G. Levi, "Automobile Fuel Efficiency: The Opportunity and the Weakness of Existing Market Incentives," *Resources and Conservation* 10 (1983): 103–24.

One technology that could cut NO_x emissions while increasing fuel economy is an oxygen-separation membrane. By separating nitrogen from air and keeping it out of the cylinders, it permits use of a smaller, lighter engine while reducing nitrogen burning. See Bleviss, *New Oil Crisis,* p. 35.

P. 126 On the all-important issue of codesign for fuel economy and low emissions, see Calwell, *Near Term Potential;* and Bleviss, *New Oil Crisis.*

P. 127 The "polynucleated urban area" is described by K.-S. Kim and J. B. Schneider, "Defining Relationships Between Urban Form and Travel Energy," in Transportation Research Board, *Energy Analysis.* On urban design, also see Percival Goodman, *The Double E* (Garden City, N.Y.: Anchor/Doubleday, 1977). Goodman draws directly and indirectly on the earlier critiques by Jane Jacobs, Lewis Mumford and others to present a coherent environmental view of urban design from the perspective of an architect. In an area generally fraught with didacticism, this gem of a book manages to be sensible, non-dogmatic and fun to read.

The Jackson quotation is from *Crabgrass Frontier,* p. 304.

Chapter 8: Eye of the Tiger

P. 131 For emissions of greenhouse gases, see G. Marland, "Fossil Fuels CO_2 Emissions: Three Countries Account for 50% in 1986," Carbon Dioxide Information Analysis Center, Oak Ridge National Laboratory, Winter 1989; "Policy Options for Stabilizing Global Climate," U.S. EPA, February 1989. For population statistics, see the International Institute for Environment and Development's *World Resources, 1988–89* (New York: Basic Books, 1988).

P. 132 Deforestation estimates are found in a paper by A. Atiq Rahman of the Bangladesh Centre for Advanced Studies, prepared for a conference,

Global Warming and Climate Change: Perspectives from Developing Countries, held in New Delhi, India, February 21–23, 1989.

Descriptions of Dhangmari derive from a visit by MO, February 1989. Discussions with K. M. Hossain and M. Hoque of the Department of Geology, University of Dacca, along with material provided by them, are the sources for geological information on Bangladesh. They also provided insights on the response to cyclones, and the resultant loss of life, as did various local officials in Khulna and Dhangmari. The first-person accounts of the villagers themselves were our main source.

P. 132 The consequences of sea-level rise for Bangladesh are discussed by J. M. Broadus, J. D. Milliman, S. F. Edwards, D. G. Aubrey, and F. Gable, "Rising Sea Level and Damming of Rivers: Possible Effects in Egypt and Bangladesh," in *Effects of Changes in Stratospheric Ozone and Global Climate,* vol. 4: *Sea Level Rise,* ed. J. G. Titus (Washington, D.C.: U.S. EPA, October 1986).

An excellent description of the personal vulnerability of Bangladeshis to sea-level rise was given by S. Mydans in "Life in Bangladesh Delta: On the Edge of Disaster," *New York Times,* June 21, 1987.

P. 135 These descriptions from a visit by MO and from discussions with A. A. Rahman and S. Huq of the Bangladesh Centre for Advanced Studies.

Khandoker's remarks were made to MO.

P. 135 The Philadelphia incinerator is discussed by M. K. Majumder in "Fresh Debate on Industrial Waste," *Dhaka Courier,* February 10–16, 1989, p. 11.

A. Rahman made this remark during the February 1989 conference in New Delhi.

P. 136 Inventories of methane emissions were provided by the U.S. EPA.

P. 136 Among the sources for descriptions of the September 1988 flood are the following articles from the *New York Times:* A. Mahmud, "Misery Rises with Rivers in Bangladesh," September 6, 1988, p. A1; "Refuge on Riverbank, Cries for Food," September 7, 1988, p. A8; and E. Sciolino, "Bangladesh Gets Pledge of Millions in Flood Relief," September 7, 1988, p. A1. Information from MO's visit supplemented their descriptions.

P. 137 Electric power and energy data on India come from J. M. Dave, "Policy Options for Developments in Response to Global Atmospheric Changes, Case Study for India for Greenhouse Effect Gases," a paper presented at a conference, Climate and Development, Hamburg, November 7–10, 1988; and from M. Dayal, R. K. Pachauri, and A. Reddy papers at the New Delhi conference, February 1989.

P. 137 On Gandhi, see J. M. Brown, *Gandhi's Rise to Power: Indian Politics*

1915–1922 (Cambridge: Judith M. Brown, 1972), p. 313. Gandhi called for two million spinning wheels, but the number adopted is unknown. Despite a short-term drop in cloth imports, the main effect of this tactic was largely symbolic.

P. 138 Descriptions of alternative energy projects derive from MO's visit to sites in February 1989, with the assistance of M. Dayal, G. D. Sootha, and B. Banerjee of the Department of Non-Conventional Energy Sources.

P. 139 Data on the United States's budget for renewable energy projects comes from D. J. Hanson and J. R. Long, "Bush Budget Proposal Calls for Increases in R&D Funding," *Chemical and Engineering News,* February 20, 1989. Data for India are from M. Dayal and M. R. Srinivasan at the New Delhi conference, February 1989.

Data on the World Bank's energy loan were assembled by P. Miller of the Environmental Defense Fund. The issue of energy efficiency in the Third World and the World Bank's lending is also addressed by J. Van Domelen in *Power to Spare: The World Bank and Electricity Conservation* (Washington, D.C.: Osborn Center, 1988).

P. 139 Critiques of the Indian renewable-energy program were supplied by G. Leach of the Stockholm Environmental Institute and Stewart Boyle of the Association for the Conservation of Energy, London.

Efficiency and renewable energy potential for India discussed by Dave in "Policy Options for Developments."

P. 141 Economic, energy, and emissions statistics for China are based on a paper by Lu Yingzhong, "The CO_2 Issues and Energy Policies in the PRC," delivered at the New Delhi conference, February 1989. Lu Yingzhong is affiliated with the Institute for Techno-economics and Energy System Analysis, Beijing. Other sources include *China: The Energy Sector* (Washington, D.C.: World Bank, 1985); the EPA study, "Policy Options"; and *World Resources, 1988–89.*

P. 141 For Marco Polo's and von Richthofen's descriptions of China's coal reserves, see T. R. Tregan, *A Geography of China* (Chicago: Aldine, 1965), pp. 145–151.

P. 142 The comparison with Japan is taken from P. M. Kennedy, *The Rise and Fall of the Great Powers* (New York: Random House, 1987).

The land requirements for wood gasification were provided by R. H. Williams.

P. 142 Our discussion of industrial energy use in China is based on *China: The Energy Sector;* on J. Goldemberg, T. B. Johansson, A. K. N. Reddy, and R. H. Williams, "An End-Use Oriented Global Energy Strategy," *Annual Review of Energy* 10 (1985): 613–88; on two reports by the same authors, "Energy for Development" and "Energy for a Sustainable

World," published in September 1987 by the World Resources Institute, Washington, D.C.; and on W. U. Chandler, "Assessing Carbon Emission Control Strategies: The Case of China," *Climatic Change* 13 (1988): 241–65.

P. 143 Data on rice fertility come from S. Yoshida, T. Satake, and D. S. Mackill, "High-Temperature Stress in Rice," International Rice Research Institute Research Paper 67 (Manila, 1981).

P. 144 Discussions of the effect of monsoon variability on agriculture are found in M. L. Parry, T. R. Carter, and N. T. Konijn, *The Impact of Climatic Variations on Agriculture* (Dordrecht, Netherlands: Kluwer Academic, 1988).

For optimistic projections of the future of agriculture, see for example P. E. Waggoner, "Agriculture and Climate Changed by More Carbon Dioxide," in *Changing Climate* (Washington, D.C.: National Academy Press, 1983). The various views are discussed in the notes to chapter 1. For dislocations related to the 1988 drought, see P. Lewis, "Rise in Hunger Seen as World's Harvests Fall and Costs Soar," *New York Times,* April 5, 1989, p. A1; see also the discussion of the relation of climatic change to food security in "State of the World, 1989," Worldwatch Institute, Washington, D.C., pp. 13–16.

P. 145 Statistics on deforestation may be found in *World Resources, 1988–89;* further information was provided by B. Rich and S. Schwartzman of the Environmental Defense Fund.

P. 146 Estimates on deforestation in the Amazon come from a paper by R. A. Houghton of the Woods Hole Research Center presented at the New Delhi conference, February 1989.

P. 147 For the anticancer properties of rain forest species, see Catherine Caufield, *In the Rainforest* (New York: Alfred A. Knopf, 1985), pp. 218–19.

See C. William Beebe, *The Bird, Its Form and Function* (New York: Henry Holt, 1906), p. 18.

Information on Costa Rica was gleaned during a visit by MO; also from *World Resources, 1988–89.*

P. 148 Information on Brazil's energy plan and the role of the World Bank was provided by B. Rich and P. Miller of the Environmental Defense Fund.

P. 148 Sustainable development was discussed by the World Commission on Environment and Development in "Our Common Future," distributed by the United Nations in 1987. Its chair was Gro Harlem Brundtland, prime minister of Norway.

P. 149 T. J. Lovejoy III, "Aid Debtor Nations' Ecology," *New York Times*, Oct. 4, 1984, A31.

P. 150 Brazil's President Sarney is quoted by M. Simons in "Brazil, Smarting from the Outcry over the Amazon, Charges Foreign Plot," *New York Times,* March 23, 1989, p. A14.

The data on forest acreage are taken from *World Resources, 1988-89.*

Timothy Egan, "Logging in Lush Alaskan Forests Profits Companies and Costs U.S.," *New York Times,* May 29, 1989, p. 1.

P. 150 Liu Ming Pu is quoted by W. K. Stevens in "Ecological Threats, Rich-Poor Tensions," *New York Times,* March 26, 1989, iv., p. 18.

P. 151 The population statistics are based on *World Resources, 1988-89.*

P. 152 An analysis of the effect of population on greenhouse-gas emissions is given in "The Full Range of Responses to Anticipated Climatic Change," United Nations Environment Programme and the Beijer Institute, April 1989. But this discussion assumes per capita energy-use growth at the same rate regardless of population size.

P. 152 Emissions for China are taken from Lu Yingzhong, "The CO_2 Issues"; population figures come from *World Resources, 1988-89.* Data for the United States were obtained from the World Resources Institute.

P. 153 A recent synopsis of population policy issues is given by N. Keyfitz in "The Growing Human Population," *Scientific American* (September 1989): 119–26. See also R. Repetto, "Population, Resource Pressures, and Poverty," in *The Global Possible,* ed. R. Repetto (New Haven: Yale University Press, 1985).

Chapter 9: The Fifth Wave

P. 156 On the recent history of the federal energy policy, see R. Cavanagh, C. Calwell, D. Goldstein, and R. Watson, "National Energy Policy," *World Policy Journal* (Spring 1989): 239–64; K. Schneider, "U.S. Spent Billions on Atom Projects That Have Failed," *New York Times,* December 12, 1988, p. A1.

P. 157 World energy-use data are based on the World Commission on Environment and Development, "Our Common Future," distributed by the United Nations in 1987. The projection of the reduction schedule needed to slow warming is based on J. Jaeger, "Developing Policies for Responding to Climatic Change" (Stockholm: Beijer Institute, April 1988); as well as the statement of a conference entitled "The Changing Atmosphere: Implications for Global Security," held in Toronto, June 30, 1988; and MO's estimates.

P. 159 On substitutes for CFCs and on the role of the Chemical Manufacturers Association, see S. L. Roan, *Ozone Crisis* (New York: John Wiley, 1989). The ICI Americas comment is from a corporate information packet called "ICI and the CFC Issue." The date that research was halted by that corporation is cited as 1981. J. Hoffman of the U.S. EPA, D. Doniger of the Natural Resources Defense Council, and D. Dudek of the Environmental Defense Fund all emphasized the significance of the Du Pont comments on the availability of substitutes, which were made originally at an EPA workshop early in 1986; see D. D. Doniger, "Politics of the Ozone Layer," *Issues in Science and Technology* 4 (1988): 86–92.

P. 159 For summaries of the range of CFC substitutes, see P. S. Zurer, "Producers, Users Grapple with the Reality of CFC Phaseout," *Chemical and Engineering News,* July 24, 1989 pp. 7–13; idem, "Search Intensifies for Alternatives to Ozone-Depleting Halocarbons," *Chemical and Engineering News,* February 8, 1988 pp. 17–20.

On events related to the Arctic findings, see P. Shabecoff, "Arctic Expedition Finds Chemical Threat to Ozone," *New York Times,* February 18, 1989, p. 1; C. R. Whitney, "London Talks Hear Call for '97 Ban on Anti-Ozone Chemical," *New York Times,* March 6, 1989, p. B10; and C. McGourty, "London Ozone Meeting Wins Some Hearts," *Nature* 338 (1989): 101.

Kenneth Arrow, "Economic Welfare and the Allocation of Resources for Invention," in *The Rate and Direction of Inventive Activity,* ed. R. Nelson (Princeton: Princeton University Press, 1962).

On the imminence of regulatory action, or lack thereof, a referendum on the ballot in California in 1990 would cut that state's carbon dioxide emissions by 40 percent, corrected for above-average population growth, over the next two decades, if approved. Also, several European governments were either committed to or were considering a freeze or modest reduction in carbon dioxide emissions. But at the U.S. federal level, regulatory action on greenhouse emissions seems far in the future.

P. 161 See Arrow, "Economic Welfare," p. 623 for quote.

P. 162 On electric-vehicle fleets, see J. J. Flink, *The Car Culture* (Cambridge, Mass.: MIT Press, 1975).

P. 163 The history of the United States Fuels Corporation (the Synfuels Corporation) was obtained from Committee on Energy and National Resources, "Synfuels from Coal and the National Synfuels Production Program: Technical, Environmental, and Economic Aspects," United States Senate, January 1981; and from MO's interview with S. Clayton of the Department of the Treasury. The Energy Security Act of 1980 established the corporation and authorized the expenditure of $88 billion in loan and price guarantees for commercial scale synthetic-fuel demonstration projects. In addition, the Department of Energy had earlier begun to support the Great Plains coal gasification project, which was never folded into the Syn-

fuels Corporation, as well as one other project in which Exxon participated, which was terminated early in the 1980s.

After the collapse of oil prices, the prospects for synfuels dimmed. Only
four projects were ever funded by the corporation, and of these, only two—
Union Oil Company's Parachute Creek oil-shale project (which has never
operated at expected levels and which produces oil at more than double the market price) and Dow Chemical's Plaquemine, Louisiana coal-
gasification project—are still in operation and drawing federal subsidies.
Since the corporation officially folded in April 1986, these remnants are
administered by the Treasury Department. In addition, the Southern California Edison "cool-water" gasification and electric-generation project has
terminated, having provided some interesting information. And another,
the Forest Hill Heavy Oil project in Texas, is in bankruptcy; while still
extracting oil, it is no longer using a synthetic-fuel process. All told, these
projects are authorized to expend about $1.2 billion before all four ter-
minate.

The Great Plains project defaulted on $1.5 billion in federal loan guar-
antees and was recently sold to the Basin Electric Power Cooperative for a
fraction of the default cost. So the federal government spent only a small
part of its original commitment to synfuels, but still managed to lose substantial sums of money and produce very little fuel.

P. 163 On the federal role in research, also see R. B. Reich, "America Pays
the Price," *New York Times Magazine,* January 29, 1989, pp. 32–40.

P. 164 Direct federal spending for research and development of renewable
energy fell by nearly 90 percent over the 1980s. Support for energy efficiency research and development fell by two-thirds. Support for fission and
fusion research and development was also slashed, but by smaller amounts.

P. 165 On fossil-fuel preferences, see A. V. Kneese, "Natural Resources
Policy 1975–1985," *Journal of Environmental Economics and Management* 3
(1976): 253–88.

P. 166 On MITI, see R. R. Nelson, "National Systems of Innovation,"
and C. Freeman, "Japan: A New National System of Innovations," in
Technical Change and Economic Theory, ed. G. Dosi, C. Freeman, R. Nelson, G. Silverberg, and L. Soete (London: Pinter, 1988), B. Harrison and
B. Bluestone, *The Great U-Turn* (New York: Basic Books, 1988); and D.
E. Sanger, "Mighty MITI Loses Its Grip," *New York Times,* July 9, 1989,
p. III-1.

P. 166 On the Competitiveness Study Group, see P. T. Kilborn, "Bush
Aides Study Ideas to Redirect Goals of Business," *New York Times,* January
9, 1989, p. A1.

P. 167 On Schumpeter and his recent disciples, see Dosi et al., eds., *Tech-*

nical Change; and J. Schumpeter, *Capitalism, Socialism, and Democracy* (London: George Allen and Unwin, 1943). There remains considerable debate over the nature and cause of long waves in economic activity. The reasons put forward vary from purely technological explanations related to innovation, to social and political factors like war, to purely economic factors like patterns of investment. The recent discussions of the neo-Schumpeterians focus on the first two of these as causes of the third. For a detailed examination of this debate, see J. S. Goldstein, *Long Cycles: Prosperity And War In The Modern Age* (New Haven, Conn.: Yale University Press, 1988). This book also contains a thorough discussion of Kondratiev's various supporters and detractors over the last 60 years. On Kondratiev's theory and biographical history see also G. Garvy, "Kondratieff's Theory of Long Cycles," *Review of Economic Statistics* 25 (1943): 203–20.

On the characterization of each economic wave, see C. Freeman and C. Perez, "Structural Crises of Adjustment, Business Cycles, and Investment Behavior," in Dosi et al., eds., *Technical Change,* chap. 3.

P. 168 The notion that certain technological, social, and industrial choices cluster together is also explored by M. J. Piore and C. F. Sabel in *The Second Industrial Divide* (New York: Basic Books, 1984). Their introduction, pp. 4 and 5, and comments on the development of the railroads, p. 67, are particularly relevant. The fact that such clusters or paradigms are self-reinforcing and tend to drive out competing clusters once they gather steam has been explored by several economists, including P. A. David and W. B. Arthur of Stanford University. This school emphasizes the significance of political and social decisions in technological history; because technology interacts with other factors such as politics, what we get in the end is not an inevitable outcome of some technological juggernaut, but depends in some measure on what we ask for. If we don't want fossil fuels, we need to start asking for solar energy and for the new industrial and social forms that will reinforce its dominance. In other words, the dominant technologies reflect specific choices, not just abstract economic and technical necessity. Isolated examples include the "Qwerty" typewriter keyboard and clockwise clocks, both of which reflect historical accidents or choices which are no longer relevant.

See K. T. Jackson, *Crabgrass Frontier: The Suburbanization of the United States* (New York: Oxford University Press, 1985), on the growth of suburbia and its relation to the car.

P. 170 See Dosi et al., eds., *Technical Change,* for the discussion of Freeman and Perez.

On the shift from wood to coal, see *Historical Statistics of the United States,* Bicentennial Edition, vol. 1 (Washington, D.C.: U.S. Department of

Commerce, 1975). Data on the shift from coal to oil and gas come from "Historic Emissions of Sulfur and Nitrogen Oxides in the United States from 1900 to 1980," document EPA-600/7-85-009a (Research Triangle Park, N.C.: U.S. EPA, April 1985).

P. 171 On U.S. and Japanese energy trends, see W. U. Chandler, H. S. Geller, and M. R. Ledbetter, *Energy Efficiency: A New Agenda* (Washington, D.C.: American Council for an Energy Efficient Economy, 1988).

P. 172 On the relation of military spending to economic growth, see "Rethinking the Military's Role in the Economy," *Technology Review* (August–September 1989): 55–64; and R. R. Nelson, "Institutions Supporting Technical Change in the United States," in Dosi et al., eds., *Technical Change,* chap. 15, which includes the quoted remark. See also the discussion of the aircraft industry in *Made in America,* ed. M. L. Dertouzos, R. K. Lester, and R. M. Solow (Cambridge, Mass.: MIT Press, 1989), pp. 11–12 and 201–16.

In addition to the global change program mentioned in chapter 4, and Mission to Planet Earth, a plan to monitor the global environment from space, other basic research programs in a variety of fields are in desperate need of support.

To comprehend fully the mismatch of need and resources, consider that federal funding for basic research in nonmilitary, nonhealth areas, including the kind of studies that will underlie almost all new civilian technologies, amounts to only $5.5 billion in the proposed 1990 budget, only 8 percent of total federal R and D obligations. If megaprojects, which provide little or nothing of importance to the survival of the planet, like the space station (which appears set to receive more than $1.5 billion in the 1990 budget) and the Superconducting Super Collider are to be funded, then small-scale basic research and vital space projects like Mission to Planet Earth will be squeezed out.

P. 173 On the Critical Technologies list, see C. Norman, "DOD Lists Critical Technologies," *Science* 243 (1989): 1543.

The quote of Sheldon Buckler was part of a personal communication to RHB.

Chapter 10: Building Blocks

P. 176 For a discussion of various regulatory approaches, particularly market-based incentives, see R. Stavins, "Project '88: Harnessing Market Forces to Protect Our Environment: Initiatives for the New President," a public policy study sponsored by senators Timothy E. Wirth and John Heinz, Washington, D.C., December 1988.

P. 177 Our fuel-economy data are taken from Calwell, "The Near Term Potential for Simultaneous Improvements in the Fuel Efficiency and Emissions of U.S. Automobiles" (Master's thesis, University of California, Berkeley, May 1989).

P. 178 On electric-utility regulation, see R. C. Cavanagh, "Least-Cost Planning Imperatives for Electric Utilities and Their Regulators," *Harvard Environmental Law Review* 10 (1986): 299–344; and D. Roe, *Dynamos and Virgins* (New York: Random House, 1984).

On household discounts, see A. H. Rosenfeld and D. Hafemeister, "Energy Efficient Buildings," *Scientific American* 258 (1988): 78–85.

P. 179 On New York State regulations see "Opinion and Order Establishing Guidelines for Bidding Program," Opinion no. 89-07 (New York Public Service Commission, April 1989).

P. 180 On the automakers' position with regard to CAFE standards and gasoline taxes, see L. Iacocca, *Talking Straight* (New York: Bantam, 1988); and P. A. Eisenstein, "Automakers Are Being Pushed Toward Fuel Economy," *Christian Science Monitor,* May 23, 1989, p. 9.

P. 180 For a discussion of the coordination of gasoline taxes and CAFE standards, see W. U. Chandler, H. S. Geller, and M. R. Ledbetter, *Energy Efficiency: A New Agenda* (Washington, D.C.: American Council for an Energy Efficient Economy, 1988). On European gasoline taxes, see D. L. Bleviss, *The New Oil Crisis and Fuel Economy Technologies* (New York: Quorum Books, 1988).

P. 182 On trees, see D. J. Dudek, *Offsetting New CO_2 Emissions* (New York: Environmental Defense Fund, 1988).

P. 184 Jessica Tuchman Mathews's remarks were made with regard to mandated methanol use at a 1988 briefing for congressional staff, in which MO participated.

P. 185 James Baker's comment is quoted by P. Shabecoff in "Joint Effort Urged to Guard Climate," *New York Times,* January 31, 1989, p. A9.

P. 185 The episode of Hansen's testimony, and the Administration's turnaround, may be traced through the following articles: M. Weisskopf, "Sununu Blocked Plan to Seek Convention on Global Warming," *Washington Post,* May 6, 1989, p. A2; P. Shabecoff, "Scientist Says Budget Office Altered His Testimony," *New York Times,* May 8, 1989, p. A1; C. Peterson, "Bush Urged to Shift Stance on Global-Warmth Conference," *Washington Post,* May 10, 1989, p. A2; M. Oppenheimer, "The Greening of Mrs. Thatcher," *New York Times,* May 10, 1989, p. A35; and P. Shabecoff, "E.P.A. Chief Says Bush Will Not Rush into a Treaty on Global Warming," *New York Times,* May 13, 1989, p. 9.

P. 187 Margaret Thatcher quote obtained from the Downing Street press office. MO was one of two Americans who participated in the briefing; he discusses its significance in "The Greening."

Chapter 11: The Last War

P. 190 On solar hot-water systems, see D. Browler, "Building with the Sun," *Metropolis* (May 1989): 71.

On federal funding for photovoltaics, see "Photovoltaics: Solar Electricity in the 1990's" (Arlington, Va.: Solar Energy Industries Association, November 1988).

On ARCO Solar, see M. L. Wald, "ARCO to Sell Siemens A.G. of West Germany Its Solar Energy Unit," *New York Times*, August 3, 1989, p. D2.

On levels of photovoltaic funding by other countries, see Solar Energy Industries Association and Solar Energy Intelligence Report, Silver Spring, Md., April 21, 1989.

The data on energy consumption are from *Historical Statistics of the United States,* Bicentennial Edition, vol. 1 (Washington, D.C.: U.S. Department of Commerce, 1975). Data on industrialization are from Paul M. Kennedy, *The Rise and Fall of the Great Powers* (New York: Random House, 1987), p. 200.

P. 193 On the German Greens and foreign policy, see S. Schmemann, "Struggling Kohl Sending 2 Aides to an Edgy U.S.," *New York Times*, April 24, 1989, p. A1.

On the Green vote, see G. R. Whitney, "Morning After Europe's Vote: Is the Landscape New in London and Bonn?" *New York Times*, June 20, 1989, p. A10.

On the Netherlands controversy, see S. Rule, "Politics as Usual over the Dutch Environment," *New York Times*, June 11, 1989, p. 10.

On the Green movement in Europe, see J. M. Markham, "Greening of European Politicians Spreads as Peril to Ecology Grows," *New York Times*, April 12, 1989, p. A1; M. Simons, "Green Parties Look for Gains in European Voting," *New York Times*, May 31, 1989, p. A3.

P. 193 On the Danube dam, see B. G. Liptak, "Austria Fouling Hungary's Environment," letter to the *New York Times*, March 9, 1988, p. A30; V. Rich, "Objectors Find a Voice," *Nature* 338 (1989): 7.

The discussion of environmentalism in the Soviet Union and Eastern Europe was informed by interviews conducted by MO with environmentalists and journalists in Leningrad and Moscow during May 1988. See also "From Below" (Washington, D.C.: Helsinki Watch, October 1987); and the excellent articles of B. Keller in the *New York Times* since 1987, such as "Storm of Protest Rages over Dam Near Leningrad," September 27, 1987, p. 16, and "Public Mistrust Curbs Soviet Nuclear Power Efforts," October 13, 1988, p. Al. Other *New York Times* stories include P. Taubman, "Estonia Nationalists Begin to Challenge Moscow Dominance," July 21, 1988, p. 1; and P. Shabecoff, "U.S. and Soviet Groups Joining for Quality of Life," December 12, 1988, p. B10.

Kennedy, *Rise and Fall.*

P. 195 On the Cold War, see J. Newhouse, *War and Peace in the Nuclear Age* (New York: Alfred A. Knopf, 1989); and W. Isaacson and E. Thomas, *The Wise Men* (New York: Simon and Schuster, 1986).

On military pork barrel, Fred Barnes, writing in the *New Republic,* quoted a White House official saying "SDI is there to be sold as a pork program. Pieces of it are produced in nearly every Congressional district" (August 21, 1989, pp. 13–14).

P. 195 Dimitri Simes is quoted from T. L. Friedman, "U.S.-Soviet Talks Turn to New Public Dangers," *New York Times,* May 5, 1989, p. A20.

P. 196 Thomas Schelling is quoted from G. Maranto, "Are We Close to the Road's End?" *Discover* (January 1986): 28–50; Schelling amplified on his remark in a conversation with MO.

P. 196 On the economic summit, see P. T. Kilborn, "Environment is Becoming Priority Issue," *New York Times,* May 15, 1989, p. D5.

Mrs. Brundtland's comment on NATO was reported on National Public Radio, May 1989.

The UN Environment Programme action refers to Intergovernmental Panel on Climate Change.

P. 197 The Highway Trust Fund data are found in the Transportation Alternatives Group's "Future Federal and Total Investment Ranges: Highway and Transit," April 7, 1989.

On decreasing military spending, see M. Renner, "National Security: The Economic and Environmental Dimension" (Washington, D.C.: Worldwatch Institute, May 1989).

P. 198 Jacques Gansler is quoted from W. J. Broad, "Military Research Facing the Pinch of Tight Budgets," *New York Times,* July 3, 1989, p. Al.

P. 199 Ivan Illich's comments are found in "The Shadow Our Future Throws," *New Perspectives Quarterly* (Spring 1989): 20–25.

P. 200 Sir Crispin Tickell is quoted from "How Warming Could Create Refugee Crisis," *The Independent* (London), June 6, 1989, p. 3.

On the Polish election and the Chinese insurrection, see the *New York Times,* June 30, 1989, p. 1.

P. 201 Daniel Cohn-Bendit is quoted from M. Simons, "Green Parties Look for Gains in European Voting," *New York Times,* May 31, 1989, p. A3.

P. 202 William K. Reilly is quoted from P. Shabecoff, "U.S. to Urge Joint Environmental Effort at Summit," *New York Times,* July 6, 1989, p. A9.

The information about Greenhouse Glasnost was obtained in a personal communication from Walter Orr Roberts to MO.

P. 203 On the economic summit, see S. R. Weisman, "Japan to Propose a Package of Aid Worth $43 Billion," *New York Times,* July 12, 1989, p. 81.

P. 204 For example, the jet engine was conceived in the late 1920s, built and operated in 1937, and used to fly an airplane in 1939. The first jet airplanes to see military service were the German Messerschmitt Me-262 and the British Gloster Meteor, both in 1944. The first commercial jet airplane, the de Havilland Comet, flew in 1952, and the Boeing 707 began service in 1958. For further details see T. K. Derry and T. I. Williams, *A Short History of Technology* (Oxford: Oxford University Press, 1960).

P. 204 On the debate over ecological stability versus economic growth, see articles in *New Perspectives Quarterly* 6 (Spring 1989): 2–47; as well as W. R. Mead, "Environmental Keynesianism," *New Perspectives Quarterly* 7 (Summer 1989): 62–63.

P. 206 R. L. Heilbroner is quoted from *An Inquiry into the Human Prospect (Updated and Reconsidered for the 1980's)* (New York: W. W. Norton, 1980), p. 154.

P. 206 See Thomas Paine, *Common Sense and Other Political Writings,* ed. Nelson F. Adkins (Indianapolis: Bobbs-Merrill, 1953), p. 51.

INDEX

Index

Baker, James, 185

Bangladesh, 131–136, 137, 144, 145

Batteries, for solar or nuclear power, 93

Becquerel, Alexandre Edmond, 87–88

Becquerel, Antoine, 88

Beebe, William, 147

Benz, Karl, 115

Biogas digesters, India and, 138–139

"Biological diversity," 73

Biomass energy, 102–103

Birds, future greenhouse effect and, 9–10, 11, 216n

Black blizzards: Dust Bowl and, 215n; future greenhouse effect and, 9

Blanchard, Thomas, 114

BMW: electric, 124; hydrogen-powered, 123

Bockris, John O'M., 92, 120

Bolin, Bert, 38, 40

Brazil, 145–150

Broecker, Wallace, 17, 82

Bromley, D. Allan, 186

Bruce, A.W., 119

Bruegel, Pieter, the Elder, 19

Brundtland, Gro Harlem, 193, 196, 249n

Brundtland Commission, 148

Bryson, Reid, 51

Buckler, Sheldon A., 173

Buick Le Sabre, 112

Bulgaria, 193

Bush, George, 57, 76, 188; acid rain and, 176–177; climate convention and, 202; energy policy, 166–167, 185; environment and, 194; *Exxon Valdez* and, 155; ozone depletion and, 50; space and, 197–198; taxation and, 180

Bush, Vannevar, 67–68

Cadillac, 114

CAFE, *see* Corporate Average Fuel Economy

Callendar, G.D., 35

Calvert, Jack, 63

Canada, acid rain and, 62

Can We Delay A Greenhouse Warming?, 39–40

Capitalism, environmentalism and, 201

Carbon, 23

Carbon dioxide (CO_2), 23

Carbon dioxide emissions: China and, 141–142; ocean absorption of, 36; reduction of, 157–158; Soviet Union and, 195

Carbon dioxide levels: atmospheric, 1, 36–37; in early 1900s, 21; fossil fuels and, 22–29; irreversible buildup of, 80; reforestation slowing buildup of, 81

Carbon monoxide: from cars, 183; from fossil fuels, 25

Caro, Robert, 117

Cars, *see* Automobiles

Carson, Rachel, 72, 73, 223n

Carter, Jimmy, 159; acid rain and, 62, 76; energy policy, 201; Synfuels Corporation and, 39, 163–164

Cattle, methane from, 32–33, 136, 153

CFCs, *see* Chlorofluorocarbons

Chamberlain, T.C., 220

"Changing Atmosphere, The," 158

Chapin, D.M., 88

Chapin, Roy D., 115

Chernobyl, 94, 100

China, 141–143, 144, 145, 150–151; cars in, 126; energy use in, 156; fossil fuels used in, 5; population growth and, 152–153

Chlorofluorocarbons (CFCs), 29, 30, 31, 42–50, 225n–227n; drainage from atmosphere, 81–82; emissions, 212n; global warming and, 38; Helsinki agreement on, 48, 50, 160; history of concern with, 42–50, 225n–227n; India and, 151; irreversible buildup of, 80; mandatory elimination of, 181; Montreal Protocol on, 18, 64, 151, 158–159, 160, 181, 186–187, 192, 193, 196, 202, 212n; nitrous oxide and, 31–32; ozone depletion and, 29–31, 40–41, 69, 158–160; substitution for, 158–160; *see also* Ozone depletion

Chronar Corporation, 97

Chrysler Corporation, 180

Cicerone, Ralph, 32

Citizens for Sensible Control of Acid Rain, 63, 230n–231n

Clark, William, 57–58

Clark, Wilson, 20

Clean Air Act of 1963, 78, 79, 84

Climate Action Network, 202

Climate feedbacks, 26, 28, 221n

Clinch River breeder reactor, 156

Clouds, acting as greenhouse gases, 28

Coal: acid rain and 62; avoiding burning, 183; carbon-dioxide levels and, 22–29; China and, 141–143; as energy source, 171; for-

Index